ESSENTIAL MATHS SKILLS FOR EXPLORING SOCIAL DATA

Sara Miller McCune founded SAGE Publishing in 1965 to support the dissemination of usable knowledge and educate a global community. SAGE publishes more than 1000 journals and over 800 new books each year, spanning a wide range of subject areas. Our growing selection of library products includes archives, data, case studies and video. SAGE remains majority owned by our founder and after her lifetime will become owned by a charitable trust that secures the company's continued independence.

Los Angeles | London | New Delhi | Singapore | Washington DC | Melbourne

ESSENTIAL MATHS SKILLS FOR EXPLORING SOCIAL DATA

RHYS C JONES

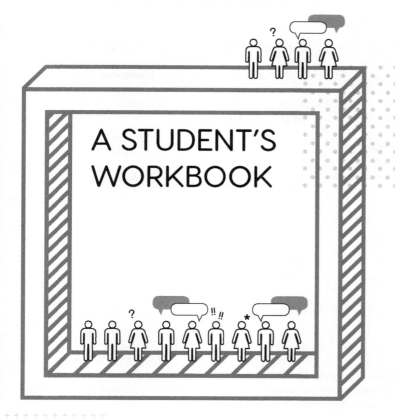

A STUDENT'S WORKBOOK

Los Angeles | London | New Delhi
Singapore | Washington DC | Melbourne

Los Angeles | London | New Delhi
Singapore | Washington DC | Melbourne

SAGE Publications Ltd
1 Oliver's Yard
55 City Road
London EC1Y 1SP

SAGE Publications Inc.
2455 Teller Road
Thousand Oaks, California 91320

SAGE Publications India Pvt Ltd
B 1/I 1 Mohan Cooperative Industrial Area
Mathura Road
New Delhi 110 044

SAGE Publications Asia-Pacific Pte Ltd
3 Church Street
#10-04 Samsung Hub
Singapore 049483

Editor: Jai Seaman
Editorial assistant: Lauren Jacobs
Assistant editor, digital: Sunita Patel
Production editor: Ian Antcliff
Copyeditor: Richard Leigh
Proofreader: Jonathan Hopkins
Marketing manager: Ben Griffin-Sherwood
Cover design: Shaun Mercier
Typeset by: C&M Digitals (P) Ltd, Chennai, India
Printed in the UK

Library of Congress Control Number: 2019950158

British Library Cataloguing in Publication data

A catalogue record for this book is available from
the British Library

ISBN 978-1-5264-6337-1
ISBN 978-1-5264-6338-8 (pbk)

CONTENTS

CONTENTS

ABOUT THE AUTHOR

Dr Rhys Christopher Jones is a Teaching Fellow in statistics based in the Department of Statistics, and also the Director of the Science Scholars Programme in the Faculty of Science, both at the University of Auckland. Originally from Cardiff in Wales, Rhys was offered the opportunity to move to Auckland, back in 2017. He took the plunge and decided to move to the other side of the planet, to the stunning shores of NZ.

Rhys teaches on an introductory course to statistics that has over 6,000 students a year at the University of Auckland, across a range of disciplines, and enjoys the challenge of bringing the subject to life and convincing people of the value of being able to analyse and interpret data confidently. He has degrees in biology, medical biochemistry, immunology, teacher training and also education, and has held lecturing positions at Cardiff University, London South East Colleges and Birmingham City University (BCU). Over his career he has taught a variety of subjects, mostly at undergraduate level, which include statistics, quantitative methods, mathematics for science, analytical and inorganic chemistry, microbiology, biomedical science, nutrition and organic chemistry, health and wellbeing, genetics, and clinical anatomy and physiology.

Rhys was also instrumental in setting up national courses in Social Analytics (the investigation of social phenomena using quantitative and statistical methods) across Wales, with the awarding body Agored Cymru. These courses were designed for students aged 16–18, to show them exciting applications of statistical techniques in areas such as the humanities and social sciences. These courses also increased the awareness of Social Analytics degrees available at Cardiff University, so that students could have a pathway to follow at higher education level.

During his time at Cardiff University, Rhys became heavily involved with setting up national training events for teachers in Wales, to help improve their teaching skills in quantitative methods. Several of these events were specifically aimed at helping teachers deliver research orientated elements and student project-based work, as part of the Welsh Baccalaureate qualification.

DISCOVER THE ONLINE RESOURCES

Get the support and guidance you need, when you need it at:
https://study.sagepub.com/jones

The website complements and builds on the material presented in the book, ensuring you have the tools you need to tackle statistical analyses with ease. It includes:

Author videos that talk you through the skills you will learn throughout the book and how they are useful, while providing practical tips and guidance on how to improve your confidence with maths and statistics.

Multiple choice questions which test your understanding of the topics covered in the text. Answers are also provided.

Flashcards to help you get to grips with key maths and statistics terminology.

Datasets and weblinks to data sources referenced in the book, so that you can better contextualise and practise your skills.

1
INTRODUCTION

You may be holding this book and thinking, 'why, oh why, do I have to study statistics? I am so bad at mathematics, it makes me so anxious, I just don't want anything to do with numbers.' You may especially be feeling like this if you are doing a humanities-based subject, sociology for example. You wanted to study sociology because you're interested in people, society, and how different groups of people interact. You did not want to look at means, standard deviations or statistical tests.

If any of the above resonates with you, don't worry – you are not alone. I have heard many students say something along these lines, even students doing science who perhaps should not be as surprised to have to do some sort of calculations. The good thing is that doing statistics at an introductory level does not actually require high-level mathematical skills. The two subjects are very different, so try not to think about them as the same thing. We live in an increasingly digitised world, data is all around us, and we are bombarded with it from all sides. Being able to talk about data, even at a descriptive level, can really help you explain a variety of social science and science based contexts.

I myself was terrible at maths in high school. When I took maths at an advanced level at age 16–18, I found it so frustrating, I just didn't get it. I'd rattle my brain for hours, getting worked up and at times really upset. When I started my under-graduate degree (biology) I knew that I'd come across maths again, and rather than get myself worked up about it I started to change my attitude. I'd have conversa-tions in my head along the lines of, 'hey, maths, I don't like you and you don't like me – but I need you in order to become a scientist'. So, I changed my attitude, which made a big difference to the maths papers I took in my first year. I ended up with A's – which I *never* thought I would achieve. And I can tell you now, this helped build my confidence to engage with the quantitative elements of my degree. Then I started doing more and more statistics, which I began to love, espe-cially the critical thinking required to ask meaningful questions about data.

My hope is that this book will help you to see the value of maths and eventually statistics, to explore data and also to understand your own discipline better.

Data is everywhere! It is an important part of everyone's lives. We see it in the news, social media, on our smart phones and in our emails. As our lives become increasingly digitised, being able to make sense of data becomes vital, not only to study the world around us, but to live in it as well! Quantitative methods in sociology, psychology, biology and other subject areas are becoming increasingly important skills to master.

This book aims to help you pick up essential mathematics skills so as to be able to explore data confidently, helping you to develop an inquisitive nature about data. The book is structured to introduce you to the chapter topics and give you an overview of what will be covered in each chapter. On this data journey, you will come across a series of case studies, which include real data from a variety of social science, biological and humanities-type data sets. As you work through the book, you should find your confidence builds as you become better able to deal with numbers. You will also develop your own critical thinking skills, which will help you to decide how to present data in different ways, along with the advantages and disadvantages of the presentations you have selected. The book is full of questions to help you get to grips with each chapter's content, and you can test yourself further by answering multiple choice questions on the accompanying website. There are also many self-directed activities, which encourage you to collect your own data and then think about the best way to present it. This is a hands-on book that was specifically created so that you can write your answers to these questions and activities in it.

Most statistics books for humanities students, especially students studying social sciences, start at too high a level. This makes them inaccessible for many students, which can add to their own anxiety. This book starts off by covering essential maths skills, and as you progress through the chapters you will notice that they build on each other. Questions will appear in later chapters that may test skills developed in previous chapters, to ensure that you are retaining these skills and practising them. This will also help you to see where everything fits together

The chapters are arranged in a similar format, starting with an outline of the skills being developed. The types of questions and resources being used in each chapter are then described, with links for you to be able to access the data sets being used. Common areas and tips to remember are included, with the case studies forming the bulk of each chapter. The questions linked to each chapter will help you build your knowledge and confidence in using statistics. Critical thinking development is also incorporated throughout the book – this is an essential skill to help you make sense of the data being presented.

Chapters 2–4 look at the essential mathematical skills required to explore data confidently. The multitude of questions included in these chapters will help you to hone your skills in being able to round numbers off to a certain number of decimal places, change between ratios, proportions and percentages, and also add, subtract, divide and multiply numbers. There is also a section on inequalities in Chapter 3, which will help you to compare two numbers and decide if one is bigger or smaller than the other.

Chapter 5 gets you to think more deeply about how tables are arranged, and about their structure and purposes. Chapter 6 introduces you to statistics, where you will begin to explore how to identify different types of data. Chapter 7 moves on to different ways of displaying data in a graph, including the key features of the main types of graph available for displaying different kinds of data. Chapter 8 begins to focus all the skills you have picked up throughout the book, and starts exploring descriptive data in more depth, helping to develop your critical thinking skills.

Chapters 9 and 10 introduce you to different ways in which we collect data, the reason why we use certain techniques, along with the advantages and disadvantages of using a variety of methods. It is envisaged that many of you will at some point in your studies have to think about creating a survey, or collecting some sort of observational data. So these chapters will get you to think about where you could take the next steps on your data exploration journey.

The data sets used in this book are in CSV or Excel format, which means you will be able to continue exploring them using other software such as SPSS, R or iNZight. Just a word of caution: some of the data sets are large, so be patient when waiting for them to download. Everything you will see in this book is sourced from real data from around the world, which may require you to think about cultural explanations that could explain the data presented. To ensure that you can keep practising your skills, there are additionally a series of online questions and activities linked to the data sets covered in the book.

Rest assured, you are in safe hands with this book. It was designed specifically for students who lack confidence in mathematics or statistics, or for students who feel they need to brush up on their skills in these areas. Good luck, and make sure you keep practising the skills you develop in this book.

Note: an initial capital letter is used to denote a variable and, also, the levels in each variable.

2

ESSENTIAL NUMERACY

Skills in this chapter

The first part of our data journey will involve developing essential numeracy skills, which will enable you to explore data more confidently. Numeracy skills are vital to analyse quantitative data (working with quantities). Essential numeracy skills include the ability to identify which values are the highest and smallest in a data set. The ability to subtract and add values together, which can change the values within a data set and change the order of the highest and smallest values, is also important. Being able to multiply and divide numbers is a core skill as well, needed especially when calculating percentages and proportions. Chapters 3 and 4 will build on using these skills, to aid in calculating percentages, ratios and proportions.

The essential numeracy skills included in this book are:

Addition: $5 + 4 = 9$

Subtraction: $5 - 4 = 1$, $4 - 5 = -1$

Multiplication: $3 \times 5 = 15$, $5 \times 0.25 = 1.25$

Division: $9/10 = 0.9$, $10/10 = 1$, $150/10 = 15$

Types of question in this chapter

This chapter presents a series of data sets, with questions aimed at developing the essential numeracy skills stated above. A series of short-answer questions accompany the case studies presented.

Resources to support this chapter

The data sets used in this chapter come from UK and New Zealand government websites (see below). Table 2.1 contains an extract from a UK data set, with several Licensing authorities removed; Table 2.2 contains the full data set. Several of the column headings in Table 2.3 were renamed to make them easier to interpret.

Case studies	Content	Theme	Country and date downloaded
2.1, 2.2	Alcohol and late night refreshment licensing, 2016	Commerce	UK, 19 April 2018

https://data.gov.uk/dataset/113b6d8d-7970-4071-b51a-0c7566786195/alcohol-and-late-night-refreshment-licensing

This data set has come from a UK government website. Access to government data has become increasingly open, with the government making claims that it is being more transparent with the data it holds. Anyone can access the data since it is completely open source. Selection of the data set was due to the interesting nature of the data. This is described in more detail in the case study descriptions.

Case study	Content	Theme	Country and date downloaded
2.3	Porirua city, dog walking areas, 2008	Environment	New Zealand, 26 June 2018

https://catalogue.data.govt.nz/dataset/dog-exercise-areas1

This data set has come from the New Zealand government website. It was selected as an example of data that the public could potentially find useful, in this case dog owners. This data set has not been updated from the original source website – this is the most up-to-date data set.

COMMON ERRORS

 It can be very easy to select the wrong numbers from a table. Ensure you look carefully at what the question is asking you to do, and that you are using the correct numbers.

 Some of the tables you come across may have missing values, which could be a dash, a zero or just left blank. This is quite common with real data. For the moment, we will just ignore these blanks and not include them in our calculations.

Local authorities and the police monitor alcohol availability, to ensure alcohol-fuelled anti-social behaviours are minimised. To support this strategy, businesses applying for alcohol licences need to go through rigorous checks before they can legally sell alcohol. Table 2.1 displays part of a table of total number of alcohol licenses Applied for in England and Wales, for the year ending 31 March 2016. Information from the table columns includes the Licensing authority and the associated Region. The numbers of New premises licences that were Applied for, Granted and Refused are also included for each Licensing authority.

Table 2.1 Alcohol licences Applied for, Granted and Refused, by type of licence, year ending 31 March 2016

Licensing authority	Region	New premises licence		
		Applied	Granted	Refused
	ENGLAND AND WALES	**9,833**	**9,064**	**271**
Aylesbury Vale	South East	18	14	0
Barnet	London	33	33	0
Barnsley	Yorkshire and The Humber	36	36	0
Barrow-in-Furness	North West	4	3	0
Basildon	East of England	17	17	0
Bath & North East Somerset	South West	52	52	0
Birmingham	West Midlands	188	182	6
Blackburn	North West	9	9	0
Blackpool	North West	53	42	5
Blaenau Gwent	Wales	4	4	0
Bolton	North West	30	30	0
Boston	East Midlands	16	16	0
Bournemouth	South West	44	40	4
Bracknell Forest	South East	9	9	0
Bradford	Yorkshire and The Humber	71	71	0

Licensing authority	Region	New premises licence		
		Applied	Granted	Refused
Braintree	East of England	18	18	0
Brent	London	61	44	1
Brentwood	East of England	8	8	0
Bridgend	Wales	18	18	0
Brighton & Hove	South East	66	57	5
Bristol	South West	143	124	9
Broadland	East of England	9	8	0
Bromley	London	54	21	–
Bromsgrove	West Midlands	13	10	0
Broxbourne	East of England	6	6	0
Broxtowe	East Midlands	12	12	0
Burnley	North West	11	11	0
Bury	North West	42	36	2
Caerphilly	Wales	23	23	0

The '–' for Bromley is an example of missing data; for the moment ignore this when answering the questions linked to this table.

Answer the following questions related to this data set.

1   Which Licensing authority had the highest number of New premises licence applications?

>

>

>

>

>

2 What is the total number of New premises license applications in Bristol and Bromley?

>

>

>

>

>

3 How many Licensing authorities come under the London region?

>

>

>

>

>

4 Which Welsh Licensing authority had the lowest number of New premises license applications?

>

>

>

>

>

5 Which Licensing authority had the highest number of New premises licenses refused?

>

>

>

>

>

6 What reason could explain the variation in alcohol licences Applied for, as seen in Table 2.1?

>

>

>

>

>

7 Bromley has missing data for the Refused column. If the Granted number of alcohol licences was actually 49 (and not 21), what would the Refused value be?

>

>

>

>

>

8 In relation to question 7, what effect would the new Refused answer have on the total number Refused for England and Wales?

>

>

>

>

>

9 If three of the Refused applications for Brighton & Hove were reconsidered and changed to Granted, what would be the new value for the number of Granted applications?

>

>

>

>

>

10 For the purposes of this data set and the accompanying report, the Licensing authorities of Bath & North East Somerset and Bristol merged. What would be the new values for the three columns of data?

>

>

>

>

>

Answers can be found at the back of the book. Test your knowledge further by answering multiple choice questions at https://study.sagepub.com/jones

Table 2.2 provides additional information on the types of premises applying for 24-hour alcohol license in England and Wales. The table includes the types of premises applying for licences, followed by the Total numbers that applied, also expressed as a percentage.

Table 2.2   Premises with 24-hour alcohol licences in England and Wales, by premises type, 31 March 2016

	Total	%
Premises with 24-hour alcohol licences	**7,029**	**100**
Pubs, bars and nightclubs	749	11
Supermarkets and stores	2,206	31
Large supermarkets	964	14
Other convenience stores	1,144	16
Supermarket and store type not reported	98	1
Hotel bars	2,978	42
Open 24 hours to residents and general public	435	6
Open 24 hours to residents and guests only	1,832	26
Hotel bar type not reported	711	10
Other premises types	977	14
Premises type not reported	119	2

Answer the following questions in relation to this data set.

11   Which premises type has the highest total 24-hour alcohol licences?

>

>

>

>

>

12   If all the Total values for Hotel bar type not reported was added to the Total values for Hotel bars open 24 hours to residents and guests only, what would the new total be?

>

>

>
>
>

13 Which premises type has the smallest percentage?

>
>
>
>
>

14 Which subdivision of premises type has the smallest percentage?

>
>
>
>
>

15 If 50 of the total Hotel bar premises registered as bars, what would the new Pubs, bars and nightclubs total become?

>
>
>
>
>

Answers can be found at the back of the book. Test your knowledge further by answering multiple choice questions at https://study.sagepub.com/jones

Public spaces can be important areas for communities to engage in outdoor pursuits – for example, children's playgrounds, parks and cycling tracks. Porirua City Council (near Wellington, New Zealand) created the data set in Table 2.3 to help inform dog owners of the public spaces available in the area, to walk their dogs. As well as the physical dimensions (length and area), a Description of the park (including its Name) is included.

Table 2.3 Porirua city, dog walking areas, 2008

Park no.	Description	Park length (m)	Name	Park area (m^2)
1	Unleashed	715.9918549	Baxters Knob Reserve (South-East Open Space)	16,000.26
2	Unleashed	328.6119424	Bedford Reserve	4,398.686
3	Unleashed	13,484.11572	Bothamley Park	806,008.8
4	Unleashed	1,217.489506	Brandon Reserve	36,790.88
5	Controlled area	168.6246326		1,190.631
6	Controlled area	743.2710457		14,232.22
7	Unleashed	2,834.199318	Camborne Walkway	33,399.57
8	Unleashed	809.5855726	Cardiff Park	31,003.47
9	Unleashed	729.7840609	Endeavour Park (southern section)	19,389.52
10	Unleashed	577.3305294	Greenmeadows Reserve	19,825.4
11	Unleashed	808.6913623	Ivey Bay Reserve	3,004.825
12	Unleashed	423.9605056	Karehana Park (southern section)	4,467.277
13	Unleashed	1,490.141576	Motukaraka Point Beach Foreshore area	15,041.99
14	Unleashed	629.6505473	Muri Reserve	18,180.81
15	Unleashed	1,757.100594	Ngatitoa Domain Foreshore area	32,669.61
16	Unleashed	304.4438508	Ngatitoa Domain (dog obedience school)	2,817.433

(Continued)

Table 2.3 (Continued)

Park no.	Description	Park length (m)	Name	Park area (m²)
17	Unleashed	1,107.797471	Onepoto Park (plantation area)	27,519.32
18	Unleashed	1,014.349719	Papakowhai Reserve	40,937.14
19	Unleashed	961.5234299	Rangituhi Park	28,290.74
20	Unleashed	1,331.669299	Spicer Botanical Park	80,214.39
21	Unleashed	3,302.43864	Spinnaker Reserve	130,608.4
22	Unleashed	4,813.123948	Staithes Scenic Reserve	170,363.1
23	Unleashed	3,146.284168	Stuart Park	251,365.2
24	Unleashed	1,084.671135	Takapuwahia Reserve	26,654.55

Answer the following questions related to this data set.

16 How many parks are described as Controlled areas?

>

>

>

>

>

17 How many parks are described as Unleashed?

>

>

>

>

>

18 Which park has the longest length?

>

>

>

>

>

19 Which park has the shortest length?

>

>

>

>

>

20 Which park has the largest area?

>

>

>

>

>

21 What is the total area of the two largest parks added together?

>

>

>

>

>

22 What is the difference between the longest and shortest parks?

>

>

>

>

>

23 Which parks have no Name? How would you describe these parks?

>

>

>

>

>

24 How many parks have an area under 1200 m²?

>

>

>

>

>

25 How many parks have a Park length over 1000 m?

>

>

>

>

>

Answers can be found at the back of the book. Test your knowledge further by answering multiple choice questions at https://study.sagepub.com/jones

TIPS TO REMEMBER

Be careful when reading from large data sets. It is very easy to mix up values, which could lead to incorrect calculations.

Pay close attention to what each question is asking you to do, to ensure you are using the correct sets of values, or that you are looking at the correct data points.

3

PERCENTAGES, DECIMAL POINTS AND INEQUALITIES

Skills in this chapter

Now that we have looked at several essential numeracy skills, we will move on to calculating percentages, rounding numbers (to a certain number of decimal places correctly), and inequalities. Every day, percentages bombard us from all sides. Sources include the TV, social media and news reports. Percentages enable the presentation of data on a scale between 0 and 100%. It is always useful to know what the sample or population size a percentage is based on. Percentages can be created from small amounts of data (a small sample) or from large amounts of data. This data may represent a whole population or only a small part. In statistics we often only have access to a sample from a bigger population of interest. For example, we might be interested in the percentage of people in Asia who prefer reading the news on a tablet, rather than in a paper newspaper. It would be practically impossible to obtain data from every individual in Asia, so we would take a sample, asking participants from the sample our question of interest. We could then calculate percentages on the sample data. Percentage values can be misleading, especially when they are generalised from a small sample to larger groups or society in general. For example, a TV advert for age-defying creams might make the claim that 87% of users agreed that their skin felt softer after 4 weeks of using a product. Then, in small print, the sample size is quoted as being 91 people. The TV advert will then suggest that you should also purchase the product, using the misleading percentage presented as a source of supporting evidence. Sampling will be covered in more detail in Chapter 10.

This chapter will develop the skills involved in calculating a percentage, as a proportion of the total number of cases relevant to the question.

To take an example, if 4 out of 10 students in a classroom were boys, as a percentage this would be:

$$\frac{4}{10} \times 100 = 40\%$$

The general formula for calculating a percentage is:

$$\frac{x}{n} \times 100$$

where x is the given quantity or amount and n is the total amount.

Another essential skill this chapter focuses on is the ability to round a number to a certain number of decimal places. Rounding to a number of decimal places, or to the nearest whole number, is stated specifically within the relevant questions linked to each case study.

Rounding to a certain number of decimal places involves paying attention to the location of the decimal point. To round numbers accurately, look at the number of decimal places stated then count that number of spaces, right from the decimal point. Then look at the number directly next to it, on the right. If it is 4 or less the number stays the same; if it is 5 or higher, you round the number up by 1. For example:

100.125

- rounded off to 1 decimal place = 100.1
- rounded off to 2 decimal places = 100.13
- to the nearest whole number = 100.

Another skill we will develop in this chapter is the use of inequality signs. Inequality signs are used to help you decide if a certain number is larger or smaller than another one. For example, if we were interested in measuring the speed of a car (for which we will use the letter s for speed, measured in miles per hour) and wanted to compare this speed to the speed limit of 30 mph in a given area, we might write this as:

$s \leq 30$

So, if our value of s was equal to or under this value, this would be legal. If the value was over this limit, it could trigger an alarm or warning (caused by the car) that the police might need to investigate.

The inequality sign used tells us whether our value of interest should be smaller than, larger than and/or equal to the value next to it. For example:

$s < 30$ means smaller than 30

$s \leq 30$ means equal to or smaller than 30

$s > 30$ means larger than 30

$s \geq 30$ means equal to or larger than 30.

Types of question in this chapter

This chapter presents a series of data sets, with questions aimed at developing skills in calculating percentages, as well as being able to perform simple rounding tasks to a stated number of decimal points. There will also be questions based on interpreting inequality statements. A series of short-answer questions accompany the case studies presented.

Resources to support this chapter

The data sets used in this chapter come from the UK, US and New Zealand government websites (see below). Table 3.1 contains anti-social behaviour order (ASBO) data for 2013, from the UK, for youth offenders. Table 3.2 contains nutritional data; it has had several columns deleted to make it easier to interpret. Table 3.3 was used in Chapter 2 and is being revisited to explore and develop skills relevant to this chapter.

Case study	Content	Theme	Country and date downloaded
3.1	ASBO statistics, 2013	Crime	UK,19 April 2018

https://data.gov.uk/dataset/25c0f3c0-e21e-4028-aaef-1981404e1825/ annual-anti-social-behaviour-order-asbo-statistics

This data set has come from a UK government website. Access to government data has become increasingly open, with the government making claims that it is being more transparent with the data it holds. Anyone can access the data since it is completely open source. News reports and many tabloids often quote crime data, being of societal importance to many people.

(Continued)

Case study	Content	Theme	Country and date downloaded
3.2	Fruit, vegetable and snack pot calories per portion, 2012	Nutrition	USA, 26 June 2018

https://www.ers.usda.gov/data-products/fruit-and-vegetable-prices.aspx

This data set was downloaded from the US Department of Agriculture. Nutritional intake, and what is 'good for us', are often contested issues in the media. Data can frequently be conflicting and confusing for people to interpret. Nutrition is an extremely important topic to look at, especially since it affects everyone – we all need food and drink to stay alive.

Case study	Content	Theme	Country and date downloaded
3.3	Porirua city, dog walking areas, 2008	Environment	New Zealand, 26 June 2018

https://catalogue.data.govt.nz/dataset/dog-exercise-areas1

This data set has come from the New Zealand government website. It was selected as an example of data that the public could potentially find useful, in this case dog owners. This data set has not been updated on the original source website – this is the most up-to-date data set.

COMMON ERRORS

 It is easy to calculate a percentage incorrectly. This may be because the numbers have been divided the wrong way around, or the answer hasn't been multiplied by 100.

 Selecting the wrong numbers will lead to an incorrect percentage calculation. Ensure you carefully read the question, so that you are choosing the right numbers to work out the percentage.

 Remember that the first step in calculating a percentage (dividing one number by another) usually results in a value smaller than 1. This step is essentially working out a proportion. To convert into a percentage (i.e. a number on a scale between 0 and 100%) you need to multiply by 100.

 Rounding to the incorrect number of decimal places can lead to inaccurate communication of your calculations. Remember to count, starting right from the decimal point, the correct number of decimal places you need to round. This is usually 2 or 3 decimal places (dp).

 Carefully check the inequality sign, and ensure you are able to ascertain which values are smaller or larger, especially if you are looking at numbers with decimal points, or comparing negative numbers. It is very easy to misread the inequality sign, or confuse negative numbers as being larger when they are actually smaller, and vice versa.

In England and Wales, anti-social behaviour covers a wide range of unacceptable activity that causes harm to individuals, communities or the environment. It could be an action by someone else that leaves an individual feeling alarmed, harassed or distressed. It also includes fear of crime or concern for public safety, public disorder or public nuisance.

Examples of anti-social behaviour include:

* nuisance, rowdy or inconsiderate neighbours
* vandalism, graffiti and fly-posting
* street drinking
* environmental damage, including littering, dumping of rubbish and abandonment of cars
* prostitution-related activity
* begging and vagrancy
* fireworks misuse
* inconsiderate or inappropriate use of vehicles.

The police, local authorities and other community safety agencies, such as Fire and Rescue and social housing property owners, all have a responsibility to deal with anti-social behaviour and to help people who are suffering from it.

Table 3.1 presents data on the numbers of offenders receiving a custodial sentence for breaching an existing anti-social behaviour order (ASBO) in England and Wales between 1 June 2000 and 31 December 2013. ASBOs were frequently issued to members of the public exhibiting anti-social behaviours in the 2000s and early 2010s. The information in the table is presented in a series of columns, starting with the length of sentence, followed by age group split up into seven different categories (10–11, 12–14, 15–17, 10–17, 18–20, 21+ and 18+). A total column represents the last set of information in this data set.

Table 3.1 Offenders receiving a custodial sentence[1] for breaching their ASBO at all courts between 1 June 2000 and 31 December 2013,[2] by Age group[3] and sentence length, in England and Wales

	Age group							Total
Length of sentence	10–11	12–14	15–17	10–17	18–20	21+	18+	
Up to and including 1 month	*	*	*	*	205	696	901	901
Over 1 month and up to 2 months	*	*	*	*	204	704	908	908

Length of sentence	Age group							Total
	10–11	12–14	15–17	10–17	18–20	21+	18+	
Over 2 months and up to 3 months	*	*	*	*	264	727	991	991
Over 3 months and up to 4 months	–	118	664	782	261	772	1,033	1,815
Over 4 months and up to 5 months	*	*	*	*	94	333	427	427
Over 5 months and up to 6 months	–	64	307	371	216	591	807	1,178
Over 6 months and up to 8 months	–	20	101	121	29	144	173	294
Over 8 months and up to 10 months	–	1	32	33	45	176	221	254
Over 10 months and up to 12 months	–	22	109	131	70	186	256	387
Over 1 year and up to 2 years	–	5	47	52	58	197	255	307
Over 2 years	*	*	*	*	10	31	41	41
Total	–	230	1,260	1,490	1,456	4,557	6,013	7,503

[1]On the occasion of the severest sentence received. Custodial sentences for breaching an ASBO may have been given concurrently with custodial sentences for other offences for which the person was found guilty.

[2]Excludes data for Cardiff Magistrates' Court for April, July and August 2008.

[3]Age is determined by age at date of appearance in court on which the severest penalty for breach of an ASBO was received.

* = missing data point

– = data still being collected at time of data entry into this table.

Answer the following questions related to this data set.

1   What percentage of the Length of sentences were up to and including a month, to the nearest whole number?

>

>

>

>

>

2 What percentage of sentences over 3 months and up to 4 months were in the Age group 10–17, the nearest whole number?

>

>

>

>

>

3 What percentage of the Length of sentences were over 1 year and up to 2 years or over 2 years, to the nearest whole number?

>

>

>

>

>

4 What percentage of the over 18 years of age sentences were 21 years of age and over, for a Length of sentence for over 2 years, to 1 decimal place (dp)?

>

>

>

>

>

5 What percentage of the 18+ Age group had sentences of up to and including 1 month or over 1 month and up to 2 months, to 2 dp?

>

>

>

>

>

6 If the 10–11 Age group total missing data point were actually 26, what would this be as a percentage of the total number of offenders for this data set, to 2 dp?

>

>

>

>

>

7 Which Age group had the second highest percentage of total offenders, and what is that percentage, to the nearest whole number?

>

>

>

>

>

8 Which Length of sentence group had the highest percentage of total offenders, and what is that percentage, to the nearest whole number?

>

>

>

>

>

9 Can you explain the general relationship or trend between age and the number of offenders receiving a custodial sentence?

>

>

>

>

>

10 Is there anything notable about the numbers of offenders as the Length of the sentence increases?

>

>

>

>

>

Answers can be found at the back of the book. Test your knowledge further by answering multiple choice questions at https://study.sagepub.com/jones

Nutritional values of common foods need to be taken into consideration to maintain a healthy diet and reduce the risks of developing diseases associated with high cholesterol or sugar consumption. Table 3.2 presents the Calories per Portion of several common types of Fruit, vegetable and Snack in the USA, in 2012. Several columns have been removed, to enable you to focus on the variables presented.

Table 3.2 Fruit, vegetable and snack pot Calories per portion, USA, 2012

Fruits and vegetables	Calories/Portion	Snacks	Calories/Portion
Apples	77	Chocolate candy	262
Applesauce, jarred	100	Cookies	123
Bananas	102	Corn chips	140
Cantaloupe	33	Crackers	114
Fruit cocktail, canned	71	Cupcakes	174
Grapes	59	Danish	271
Oranges, navel	53	Donuts	235
Peaches, canned	68	Fruit rolls	82
Pineapple, canned	75	Graham crackers	102
Plums	38	Granola bars	119
Raisins	109	Ice cream	196
Strawberries	27	Muffins	369
Tangerines	72	Pizza, from frozen	252
Watermelon	74	Popsicles and bars	80
Broccoli florets	12	Potato chips	169
Carrots, baby	22	Pretzels	168
Celery	10	Pudding, ready to eat	152
Red peppers	23	Sandwich crackers	183
Sweet potatoes, cooked	90	Toaster pastries	299
Tomatoes, grape/cherry	16	Tortilla chips	161

CASE STUDY 3.2

Answer the following questions related to this data set (to 2 dp where applicable).

11 If you decided to have fruit salad with a portion of strawberries, plums and peaches (canned), what would be the percentage of Calories of the plums in relation to the total Calories for the fruit salad?

>

>

>

>

>

12 In relation to question 11, how would your answer change if apples were added to the fruit salad?

>

>

>

>

>

13 Which food types have a higher level of Calories/Portion? fruits and vegetables or snacks?

>

>

>

>

>

14 If you were planning a party and decided to provide 5 Portions of potato chips, 10 Portions of pretzels and 7 Portions of tortilla chips, what percentage of the total Calories would the potato chips account for?

>

>

>

>

>

15 In relation to question 14, how would your answer change if the Portion of pretzels were reduced by 50%?

>

>

>

>

>

Answers can be found at the back of the book. Test your knowledge further by answering multiple choice questions at https://study.sagepub.com/jones

You have already seen this case study, in Chapter 2. This data set is being revisited to explore and develop your skills calculating percentages, rounding off numbers and interpreting inequality statements.

Public spaces can be important areas for communities to engage in outdoor pursuits – for example, children's playgrounds, parks and cycling tracks. Porirua City Council (near Wellington, New Zealand) created the data set in Table 3.3 to help inform dog owners of the public spaces available in the area, to walk their dogs. As well as the physical dimensions (length and area), a Description of the park (including its Name) is included.

Table 3.3   Porirua city, dog walking areas, 2008

Park no.	Description	Park length (m)	Name	Park area (m²)
1	Unleased	715.9918549	Baxters Knob Reserve (south-east open space)	16,000.26
2	Unleased	328.6119424	Bedford Reserve	4,398.686
3	Unleased	13484.11572	Bothamley Park	806,008.8
4	Unleased	1217.489506	Brandon Reserve	36,790.88
5	Controlled area	168.6246326		1,190.631
6	Controlled area	743.2710457		14,232.22
7	Unleased	2834.199318	Camborne Walkway	33,399.57
8	Unleased	809.5855726	Cardiff Park	31,003.47
9	Unleased	729.7840609	Endeavour Park (southern section)	19,389.52
10	Unleased	577.3305294	Greenmeadows Reserve	19,825.4
11	Unleased	808.6913623	Ivey Bay Reserve	3,004.825
12	Unleased	423.9605056	Karehana Park (southern section)	4,467.277
13	Unleased	1490.141576	Motukaraka Point Beach Foreshore Area	15,041.99
14	Unleased	629.6505473	Muri Reserve	18,180.81

Park no.	Description	Park length (m)	Name	Park area (m^2)
15	Unleased	1757.100594	Ngatitoa Domain (Foreshore area)	32,669.61
16	Unleased	304.4438508	Ngatitoa Domain (dog obedience school)	2,817.433
17	Unleased	1107.797471	Onepoto Park (plantation area)	27,519.32
18	Unleased	1014.349719	Papakowhai Reserve	40,937.14
19	Unleased	961.5234299	Rangituhi Park	28,290.74
20	Unleased	1331.669299	Spicer Botanical Park	80,214.39
21	Unleased	3302.43864	Spinnaker Reserve	130,608.4
22	Unleased	4813.123948	Staithes Scenic Reserve	170,363.1
23	Unleased	3146.284168	Stuart Park	251,365.2
24	Unleased	1084.671135	Takapuwahia Reserve	26,654.55

Answer the following questions related to this data set.

16 What is the area of the smallest park, to 2 dp?

>

>

>

>

>

17 What percentage of the parks have no names in the data set, to 3 dp?

>

>

>

>

>

18 What percentage of the parks have the word Park in their title to 3 dp?

>

>

>

>

>

19 What percentage of the parks have the word Reserve in their name to 3 dp?

>

>

>

>

>

20 What is the combined percentage of your answers to questions 18 and 19, to 2 dp?

>

>

>

>

>

21 If we wanted to check out which parks in Porirua city were over 80,000 m², and we used the letter *P* to denote this interest, how would we write this using an inequality sign?

>

>

>

>

>

22 How many parks would be included in your answer for question 21?

>

>

>

>

>

23 How would your inequality sign change for question 21 if we also wanted to include parks that had an area of 80,000 m²?

>

>

>

>

>

24 Look at the following inequality and decide if the values next to it agree with what we are looking for (for values of p). The first one has been done for you:

>

>

>

>

>

$p \leq 0.05$

0.059 (no, since this value is larger than 0.05)

0.12

0.8

1.5

0.05

1.7

0.03

25 If the inequality sign for question 24 was changed to <, would any of your answers change?

>

>

>

>

>

Answers can be found at the back of the book. Test your knowledge further by answering multiple choice questions at https://study.sagepub.com/jones

....TIPS TO REMEMBER

Pay close attention to what each question is asking you to do, to ensure you are using the correct sets of values, or that you are looking at the correct data points.

Percentages and proportions are interchangeable; for example, 70 out of 387 can be presented as a percentage, $(70/387) \times 100 = 18.1\%$ to 1 dp, or as a proportion, $70/387 = 0.2$ to 1 dp. The difference is that percentages fall on a scale between 0 and 100%, and proportions are on a scale between 0 and 1. Questions based on proportions are addressed in Chapter 4.

Make sure you carefully check the inequality sign, and ensure you are able to ascertain which values are smaller or larger, especially if you are looking at numbers with decimal points, or comparing negative numbers.

4

RATIOS AND PROPORTIONS

Skills in this chapter

It is important to be able to present information in different ways, especially when you are trying to communicate or tell a story with data. The ability to display data as ratios or proportions is an important skill that we will cover in this chapter. A ratio is a relationship between two numbers indicating how many times the first number contains the second. For example, if 20 people were asked to vote for one of two candidates, with 15 voting for the first candidate and 5 voting for the second, as a ratio this would be written as 3:1. Ratios and proportions are interchangeable, as are percentages and proportions (Chapter 3). In the example above, using percentages you could say 75% voted for the first candidate, while 25% voted for the second. Alternatively, as a proportion, you could say $\frac{3}{4}$ (simplified from 15/20) voted for the first candidate while $\frac{1}{4}$ (simplified from 5/20) voted for the second. These are just different ways to display the same data.

Types of question in this chapter

This chapter presents a series of data sets, with questions aimed at developing skills in calculating ratios and proportions. A series of short-answer style questions accompany the case studies presented. This chapter will also test skills gained in previous chapters, to ensure they are built upon and reinforced.

Resources to support this chapter

The data sets used in this chapter come from the UK and US government websites (see below). Table 4.1 contains chronic disease indicators data, collected during the period 2001–2016 and reduced to two columns, from the USA. Table 4.2 contains obesity rate data from the UK. Table 4.3 contains interesting sociological data exploring the percentage of people who feel they belong to their neighbourhood in the UK, in 2010. This data set has been shortened so it can be included in this chapter.

Case study	Content	Theme	Country and date downloaded
4.1	US chronic disease indicators, 2001–2016	Health/medical	USA, 6 June 2018

https://catalog.data.gov/dataset/u-s-chronic-disease-indicators-cdi
Associated information: https://www.cdc.gov/mmwr/pdf/rr/rr6401.pdf

This data set has come from the Center for Disease Control and Prevention, based in the USA. Chronic disease levels vary from country to country, and give a good indication of regional health needs in the population. This is described in more detail in the case study descriptions.

Case study	Content	Theme	Country and date downloaded
4.2	Obesity rates (%) among primary school children in Year 6, UK, 2006–2009	Public health	UK, 20 June 2018

https://data.gov.uk/dataset/36ba5cac-cc4e-42b2-b3e1-984daccf3c3a/ni-056-obesity-in-primary-school-age-children-in-year-6

This data set has come from a UK government website. The health of young children is of central importance to parents, schools and the government. They represent the future of a nation, and as such need to be checked for changes in health, such as obesity rates.

Case study	Content	Theme	Country and date downloaded
4.3	Percentage of people who feel they belong to their neighbourhood, 2010	Public health	UK, 20 June 2018

https://data.gov.uk/dataset/fca0e797-4bef-4983-9689-aec792f29898/ni-002-percentage-of-people-who-feel-that-they-belong-to-their-neighbourhood

This data set has come from a UK government website. Social cohesion, and feeling a sense of belonging to the neighborhoods people live in, are important contributers to health and well being for individuals and groups of people.

COMMON ERRORS

It is easy to calculate a ratio or proportion incorrectly. This may be because the numbers have been divided the wrong way around, or the numbers haven't been put in the correct order.

Selecting the wrong numbers will lead to an incorrect ratio or proportion. Ensure you carefully read the question, so that you are choosing the right numbers.

Medical data is an extremely useful source of evidence to help assess the health of a region or nation. Table 4.1 provides you with details of the Location and Disease reported of patients from across the USA. The original data set contains an extensive array of health indicators and other variables of interest useful to government officials and health-care providers. The original data set is a lot bigger, compared to the short version presented in this book. Look at the original data set and see if you can make sense of the additional information provided. This is real data collected from hundreds of thousands of patients across the USA.

Table 4.1 US chronic disease indicators, 2001–2016

Location	Disease reported
Colorado	Asthma
Washington	Asthma
New Hampshire	Diabetes
Illinois	Overarching conditions
Hawaii	Overarching conditions
South Dakota	Cardiovascular disease
South Dakota	Cardiovascular disease
Washington	Overarching conditions
Maryland	Overarching conditions
Pennsylvania	Overarching conditions
Colorado	Diabetes
Colorado	Diabetes
Washington	Asthma
Connecticut	Overarching conditions
Colorado	Asthma
Washington	Asthma
Delaware	Cancer
Wisconsin	Cardiovascular disease
New York	Overarching conditions
Virginia	Overarching conditions

Answer the following questions on the data set in Table 4.1.

1 How many patients have Overarching conditions?

>

>

>

>

>

2 What is the proportion of patients with Overarching conditions, compared to all diseases reported in Table 4.1?

>

>

>

>

>

3 How many patients were recorded from South Dakota?

>

>

>

>

>

4 What is the proportion of patients from South Dakota, compared to all Locations reported in Table 4.1?

>

>

>

>

>

5 What is the ratio of patients with Asthma, compared to all Diseases reported?

>

>

>

>

>

Answers can be found at the back of the book. Test your knowledge further by answering multiple choice questions at https://study.sagepub.com/jones

Obesity rates among young children in the UK are of growing concern and continue to rise year on year. Table 4.2 has the most up-to-date data available from the data.gov website, which presents the percentage of 10–11-year-olds who are obese in different locations across the UK. The data set presents obesity rates across three years, from 2006 to 2009. It has been reduced in size, to enable you to focus more on developing your skills in calculating proportions and ratios.

Table 4.2 Obesity rates (%) for Year 6 (ages 10–11), UK, 2006–2009

Authority	2006/07	2007/08	2008/09
Southwark	27.0	26.0	26.7
Tower Hamlets	23.0	24.5	25.7
Lambeth	25.1	23.2	25.3
Newham	23.6	25.6	24.6
Sandwell	20.2	23.9	24.6
Barking and Dagenham	20.8	23.9	24.2
Hackney	24.4	23.6	24.0
Wolverhampton	25.5	22.1	23.8
Westminster, City of	21.8	24.8	23.6
Hounslow	21.8	22.7	23.5
Knowsley	18.1	21.0	23.3
Enfield	21.4	22.5	23.0
Brent	22.1	22.5	22.9
Greenwich	21.2	22.6	22.9
Gateshead	20.2	21.6	22.8
Hartlepool	24.2	25.6	22.8
Manchester	22.8	21.9	22.6
Liverpool	18.0	20.8	22.6

(Continued)

Table 4.2   (Continued)

Authority	2006/07	2007/08	2008/09
Nottingham	20.1	22.0	22.6
Hammersmith and Fulham	23.2	22.8	22.4
Kensington and Chelsea	21.5	20.7	22.4
Halton	22.4	21.8	22.2
Lewisham	19.5	25.3	22.1
Ealing	21.8	21.0	21.9
Newcastle upon Tyne	21.3	20.8	21.9

Answer the following questions related to this data set.

6   Assume the sample size for Manchester is 50,000 for each year, and for Newcastle upon Tyne is 45,000 for each year. Draw a smaller table with just these two authorities and round these percentages to the nearest whole numbers. Your table should include values for the three years of obesity rates recorded for each Authority.

>

>

>

>

>

7   For the Manchester Authority, what is the proportion of obese 10–11-year-olds for 2007/08, in relation to the total number of obese 10–11-year-olds for 2006–2009?

>

>

>

>

>

8 What is the total number of 10–11-year-olds who were obese, from 2006 to 2009, for the Newcastle upon Tyne Authority?

>

>

>

>

>

9 For the Newcastle upon Tyne Authority, what is the proportion of obese 10–11-year-olds for 2007/08, in relation to the total number of obese 10–11-year-olds for 2006–2009?

>

>

>

>

>

10 The proportion of obese 10–11-year-olds in Manchester, for 2008/09, out of the total obese 10–11-year-olds for 2006–2009, is 9855/28,800. This can be simplified to 219/640. Round this fraction to the nearest 100th, then simplify. Finally, convert your answer into a ratio.

>

>

>

>

>

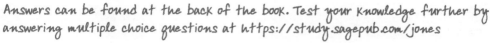

Answers can be found at the back of the book. Test your knowledge further by answering multiple choice questions at https://study.sagepub.com/jones

The health and well-being of a nation include important indicators that can influence resource allocation for publicly funded services in the UK, such as the National Health Service. Social cohesion can also have an impact on the health and well-being of a community, which in turn can be influenced by people within a community or neighbourhood feeling a sense of belonging. This sense of belonging can help people to identify themselves as being part of a community or neighbourhood, which can have a positive impact on their well-being. Table 4.3 presents data from different London boroughs, from a sample of people asked in 2008 whether they felt they belonged to their neighbourhood. The table is a smaller version of the original data set.

Table 4.3 Percentage of people who feel that they belong to their neighbourhood, London, 2010

Authority	Percentage of people
Barking and Dagenham	44.8
Barnet	56.8
Bexley	54.5
Brent	48.9
Bromley	57.4
Camden	51.9
City of London	58.7
Croydon	51.1
Ealing	49.7
Enfield	51.6
Greenwich	50.0
Hackney	57.3
Hammersmith and Fulham	49.6
Haringey	50.8
Harrow	54.3
Havering	56.7
Hillingdon	52.3
Hounslow	52.3
Islington	50.5
Kensington and Chelsea	63.7

Authority	Percentage of people
Kingston upon Thames	50.4
Lambeth	49.2
Lewisham	48.8
Merton	51.7
Newham	47.4
Redbridge	52.4
Richmond upon Thames	64.1
Southwark	48.9
Sutton	53.5

Answer the following questions related to this data set.

11 Which Authorities in Table 4.3 had the highest and lowest Percentage of people feeling that they belonged to their neighbourhood?

>

>

>

>

>

12 What is the difference between the values in your answer to question 11?

>

>

>

>

>

13 Assume that a sample of 70,000 people were asked for each of the Authorities in your answer to question 11. Calculate the number of people who felt they belonged to their neighbourhood.

>

>

>

>

>

14 Calculate the number of people who did not feel they belonged to their neighbourhood, using your answers from question 13.

>

>

>

>

>

15 Construct a table to summarise your answers for questions 13 and 14.

>

>

>

>

>

16 What is the proportion of people in Barking and Dagenham who felt they did not belong to their neighbourhood in your table constructed for question 15?

>

>

>

>

>

17 What is the proportion of people in Richmond upon Thames who felt they did belong to their neighbourhood in your table constructed for question 15?

>

>

>

>

>

18 What is the ratio of people in Barking and Dagenham who felt they did not belong to their neighbourhood, versus those who felt they did belong, in your table constructed for question 15?

>

>

>

>

>

19 What is the ratio of people in Richmond upon Thames who felt they did belong to their neighbourhood, versus those who did not feel they belonged, in your table constructed for question 15?

>

>

>

>

>

20 Why can percentages be misleading? What is the advantage of presenting data using a proportion?

>

>

>

>

>

Answers can be found at the back of the book. Test your knowledge further by answering multiple choice questions at https://study.sagepub.com/jones

TIPS TO REMEMBER

 Be careful when reading from large data sets. It is very easy to mix up values, which could lead to incorrect answers.

 Pay close attention to what each question is asking you to do, to ensure you are using the correct sets of values, or that you are looking at the correct data points.

5

TABLES

Skills in this chapter

Most of the data you have come across in this book have come in a table format. Now we will explore advanced skills in reading, interpreting and commenting on data presented in this way. Data is often presented in tables to highlight interesting patterns or relationships. The ability to read tables, spot patterns of interest, and then communicate those findings is an extremely valuable skill applicable across a range of disciplines. This chapter will build on previous chapters, as well as guiding you in constructing tables of your own from larger data sets.

Types of question in this chapter

This chapter presents a series of data sets, with questions aimed at developing skills in reading and interpreting tables, as well as producing them to summarise data. A series of short-answer-style questions accompany the case studies presented.

Resources to support this chapter

The data sets used in this chapter come from the Australian, UK, New Zealand and US government websites (see below). Table 5.1 contains annual population data relating to 2012–2017, reduced to include fewer States, from the USA. Tables 5.2 contains area and length data of sports grounds in a New Zealand town in 2008. Table 5.3 contains inspection data for farms that produce milk, relating to 2018. This data set has been shortened so it can be included in this chapter. Table 5.4

contains crime data for the state of Victoria, Australia, in 2016. Reduction in size of the data set will enable you to develop your skills in reading tables, as well as extracting important pieces of information from them.

Case study	Content	Theme	Country and date downloaded
5.1	Annual population, 2012–2017	Demographics	USA, 9 July 2018

https://factfinder.census.gov/faces/tableservices/jsf/pages/productview. xhtml?pid=PEP_2017_PEPANNRES&src=pt

This data set comes from the US Census Bureau. The data set was chosen to present the reader with a source of information that can help inform many high level decisions for policy makers, government and also businesses. Changes to population levels within countries can have a profound effect on resource allocation, for example.

Case study	Content	Theme	Country and date downloaded
5.2	Porirua city sports grounds, 2008	Environment	New Zealand, 26 June 2018

https://catalogue.data.govt.nz/dataset/sportsgrounds

This data set comes from a New Zealand government website. Data like this could be used by families and other people to look for suitability sized parks with sports facilities, near to them.

Case study	Content	Theme	Country and date downloaded
5.3	Raw drinking milk premises in England, Wales and Northern Ireland, 1 April 2018	Agriculture	UK, 1 May 2018

https://data.gov.uk/dataset/f6706084-9c82-4a50-a781-41e0e6229948/raw-drinking-milk-premises-in-england-wales-and-northern-ireland

This data set comes from a UK government website. Milk is a vital source of nutrition used in most households in the UK. Milk availability is often in the news, especially in terms of costs and profit margins for farmers. This data set presents you with a series of farms, along with the type of animal present at each farm.

Case study	Content	Theme	Country and date downloaded
5.4	Police region crime statistics, Victoria, Australia, 2016	Crime	Australia, 18 October 2018

https://search.data.gov.au/dataset/ds-dga-1ba111da-0eeb-4736-98ab-7b3f5c7a6bbd/details?q=crime

This data set comes from an Australian government website. Crime data are useful to policy makers, as well as individuals responsible for resource allocation to pay for extra police officers, for example. More information about this data set is presented in the associated case study.

COMMON ERRORS

 Be careful when reading from large data sets. It is very easy to mix up values, which could lead to incorrect calculations.

 Pay close attention to what each question is asking you to do, to ensure you are using the correct sets of values, or that you are looking at the correct data points.

Population data is an extremely important source of information, for many different sectors and research areas. For example, governments need to know current and projected population levels to help with resource allocation, businesses need to be kept up to date with population changes, which could influence the location of retail outlet stores or distribution centres. Table 5.1 displays the estimated populations of a large number of US States from 2012 to 2017. This table is a shorter version of the online data set.

Table 5.1 US annual population, 2012–2017 (population estimates, 1 July)

State	2012	2013	2014	2015	2016	2017
Alabama	4,813,946	4,827,660	4,840,037	4,850,858	4,860,545	4,874,747
Alaska	730,825	736,760	736,759	737,979	741,522	739,795
Arizona	6,544,211	6,616,124	6,706,435	6,802,262	6,908,642	7,016,270
Arkansas	2,949,208	2,956,780	2,964,800	2,975,626	2,988,231	3,004,279
California	38,019,006	38,347,383	38,701,278	39,032,444	39,296,476	39,536,653
Colorado	5,186,330	5,262,556	5,342,311	5,440,445	5,530,105	5,607,154
Connecticut	3,597,705	3,602,470	3,600,188	3,593,862	3,587,685	3,588,184
Delaware	916,868	925,114	934,805	944,107	952,698	961,939
District of Columbia	635,630	650,114	660,797	672,736	684,336	693,972
Florida	19,341,327	19,584,927	19,897,747	20,268,567	20,656,589	20,984,400
Georgia	9,911,171	9,981,773	10,083,850	10,199,533	10,313,620	10,429,379
Hawaii	1,392,772	1,408,038	1,417,710	1,426,320	1,428,683	1,427,538
Idaho	1,594,673	1,610,187	1,630,391	1,649,324	1,680,026	1,716,943
Illinois	12,878,494	12,890,403	12,882,438	12,862,051	12,835,726	12,802,023
Indiana	6,535,665	6,567,484	6,593,182	6,610,596	6,634,007	6,666,818
Iowa	3,074,386	3,089,876	3,105,563	3,118,473	3,130,869	3,145,711
Kansas	2,885,316	2,892,900	2,899,553	2,905,789	2,907,731	2,913,123
Kentucky	4,383,673	4,399,121	4,410,415	4,422,057	4,436,113	4,454,189
Louisiana	4,602,681	4,626,795	4,648,797	4,671,211	4,686,157	4,684,333
Maine	1,328,101	1,327,975	1,328,903	1,327,787	1,330,232	1,335,907
Maryland	5,891,680	5,932,654	5,970,245	6,000,561	6,024,752	6,052,177
Massachusetts	6,659,627	6,711,138	6,757,925	6,794,002	6,823,721	6,859,819

State	2012	2013	2014	2015	2016	2017
Michigan	9,886,610	9,899,219	9,914,675	9,918,170	9,933,445	9,962,311
Minnesota	5,377,695	5,416,074	5,452,649	5,483,238	5,525,050	5,576,606
Mississippi	2,982,963	2,987,721	2,988,578	2,985,297	2,985,415	2,984,100
Missouri	6,023,267	6,041,142	6,058,014	6,072,640	6,091,176	6,113,532
Montana	1,003,522	1,011,921	1,019,931	1,028,317	1,038,656	1,050,493
Nebraska	1,854,862	1,867,414	1,880,920	1,893,564	1,907,603	1,920,076
Nevada	2,752,410	2,786,547	2,831,730	2,883,057	2,939,254	2,998,039
New Hampshire	1,320,923	1,322,622	1,328,684	1,330,134	1,335,015	1,342,795
New Jersey	8,882,095	8,913,735	8,943,010	8,960,001	8,978,416	9,005,644
New Mexico	2,083,590	2,085,161	2,083,207	2,082,264	2,085,432	2,088,070
New York	19,625,409	19,712,514	19,773,580	19,819,347	19,836,286	19,849,399

Answer the following questions related to this data set.

1 All States saw an increase in population from 2016 to 2017. True or false?

>

>

>

>

>

2 Which State had the highest population in 2012?

>

>

>

>

>

3 The District of Columbia had the smallest population in 2012. True or false?

>

>

>

>

>

4 For Nevada, assume that the population for each year is comprised of 55% females and 45% males. Construct a table to show the numbers of males and females in Nevada from 2012 to 2017. Do not round any numbers in your table.

>

>

>

>

>

5 What is the advantage of looking at each individual US State's population, rather than looking at the whole country's population?

>

>

>

>

>

6 What impact does the size of a State have on resource allocation and on representation in elections?

>

>

>

>

>

Answers can be found at the back of the book. Test your knowledge further by answering multiple choice questions at https://study.sagepub.com/jones

Recreational facilities are extremely important for the public to enjoy, usually located in urban communities. Table 5.2 contains information about the Length and Area of sports grounds in the city of Porirua (near Wellington), New Zealand. Porirua City Council created this data set to help inform the public of the spaces available, to help them identify areas to engage in sports.

Table 5.2   Porirua sports grounds, New Zealand, 2008

Name of ground	Length (m)	Area (m²)
Porirua Park	2,149.166	94,026.57
Ascot Park	1,217.008	83,892.3
Adventure Park	735.6938	34,219.91
Cannons Creek Park	1,733.668	92,754.57
Cardiff Park	738.2273	21,771.65
Elsdon Park	972.016	34,164.66
Endeavour Park	841.9988	40,569.1
Kura Park	631.7027	24,884.28
Mungavin Park Netball Courts	887.1812	21,882.33
Natone Park	914.9136	29,085.06
Ngatitoa Domain	2,108.821	130,707.1
Onepoto Park	1,374.86	56,956.38
Postgate Park	501.1809	12,643.92
Rangituhi Park	762.0076	18,737.74
Whitby Tennis Courts	371.5271	6,880.55
Jillet Park Tennis Courts	447.2301	4,588.244
Waitangirua Park Courts	166.3815	1,304.693
Lagden Reserve Tennis Court	102.1586	606.1889
Makora Reserve Tennis Court	438.2563	4,745.051
Pukerua Bay Tennis Courts	216.6464	3,001.165

(Continued)

Table 5.2 (Continued)

Name of ground	Length (m)	Area (m²)
Cannons Creek Basketball Court	143.5841	1,096.572
Waihora Park	658.7816	20,218.09
Plimmerton Domain	1,756.546	73,765.33
Plimmerton Tennis Courts	235.3812	3,496.231

Answer the following questions related to this data set.

7   How could you improve the layout of Table 5.2?

>

>

>

>

>

8   If you were a resident in Porirua considering which park you would like to walk your dog in, what other pieces of information might be useful (other than what is provided in Table 5.2) to help you make your choice?

>

>

>

>

>

Answers can be found at the back of the book. Test your knowledge further by answering multiple choice questions at https://study.sagepub.com/jones

The agricultural industry in the UK provides the majority of milk supplies to supermarkets and ships across the country. Raw milk has seen growing demand from consumers, which usually means the milk has not gone through a pasteurisation process. To ensure the milk is safe to drink, all farms that produce raw milk need to undergo an inspection to check for safety compliance. Farms are graded as 'Good', 'Generally satisfactory' or 'Improvement necessary' (see Table 5.3).

Table 5.3   Farms producing raw drinking milk, England, Wales and Northern Ireland, 2018

Farm address 1	Farm address 2	Has cows	Has sheep	Has goats	Compliance rating
Lodge Farm	Newport Pagnell	Yes			Good
North Hill Farm	Buckingham	Yes			Good
Home Farm	West Hartlepool	Yes			Good
Street Farm	Chester	Yes			Good
Manor Farm	Tarporley	Yes			Good
Addashaw Farm	Middlewich	Yes			Good
Hoolgrave Manor	Crewe	Yes			Good
Home Farm	Northwich	Yes			Good
Tremedda	St Ives	Yes			Good
Seaville Farm	Wigton	Yes			Good
Slack House Farm	Brampton	Yes			Generally satisfactory
Blue Dial Farm	Maryport	Yes			Good
Low Sizergh Farm	Kendal	Yes			Good
Whinyeats Farm	Carnforth	Yes			Good
Manor Farm	Broughton in Furness	Yes			Good
Beltonville Farm	Buxton	Yes			Good

(Continued)

Table 5.3 (Continued)

Farm address 1	Farm address 2	Has cows	Has sheep	Has goats	Compliance rating
Manor Farm	Chesterfield	Yes			Good
Woodside Farm	Swadlincote	Yes			Good
West Middlewick Farm	Tiverton	Yes			Good
Modbury Farm	Bridport	Yes			Good
Park Farm	Blandford Forum	Yes			Good
Middle Farm	Sturminster Newton	Yes			Good
Meadow Cottage	Lower Hearn	Yes			Good
Northney Farm	Havant	Yes			Good
Peake House Farm	Whitchurch	Yes			Good
Westridge Farm		Yes			Good
Beacon Hill Farm	Ledbury	Yes			Good
Carpenters Hill Farm	Redditch	Yes			Good
Rumblers Farm	Hemel Hempstead	Yes			Good
Off Green Glade	Epping		Yes	Yes	Good
Redlays Farm	Hastings	Yes			Improvement necessary

Answer the following questions related to this data set.

9 Are there any ways the data in the Compliance rating column could be changed to make it easier to read? Are there any other ways to present this data?

>

>

>

>

>

10 What are the advantages of using the alternative forms you mentioned in your answer to question 9?

>

>

>

>

>

11 How do you think the farms could use the data in Table 5.3?

>

>

>

>

>

Answers can be found at the back of the book. Test your knowledge further by answering multiple choice questions at https://study.sagepub.com/jones

Crime data is becoming increasingly available from law enforcement agencies, enabling both the public and the press to look at changes in crime levels over time. Table 5.4 presents a list of types of crimes committed in the state of Victoria in Australia, showing previous (from the year before) and Current levels of each type of crime. A Percentage change in crime levels is also presented.

Table 5.4 Police regional crime statistics, Victoria, Australia, 2016

Region	Label	Previous	Current	Change (%)
1	01_Crimes against the Person	1,124.4	1,145.7	1.9
1	02_Crimes against Property	9,182.2	8,184.6	–10.9
1	03_Drug Offences	511.6	497.1	–2.8
1	04_Other Offences	1,012.1	1,394.3	37.8
1	05_Total Crime	11,830.4	11,221.7	–5.1
1	06_Robbery	146.8	130.3	–11.2
1	07_Assault	825.7	875.9	6.1
1	08_Property Damage	1,125.8	1,258.3	11.8
1	09_Burglary (Residential)	760.5	625.4	–17.8
1	10_Burglary (Other)	594.9	433.7	–27.1
1	11_Theft from Motor Vehicles	2,101.3	1,545.1	–26.5
1	12_Theft of Motor Vehicles	466.5	367.6	–21.2
1	13_Fatalities	3.8	3	–20
1	14_Serious Injuries	183	171.1	–6.5
2	01_Crimes against the Person	828.7	843.2	1.8
2	02_Crimes against Property	5,527.2	5,294.8	–4.2
2	03_Drug Offences	305.1	316.2	3.6
2	04_Other Offences	699.5	788.7	12.7
2	05_Total Crime	7,360.5	7,242.9	–1.6
2	06_Robbery	59.3	60.5	2
2	07_Assault	609.8	642.3	5.3
2	08_Property Damage	1,123.2	1,144.2	1.9

Region	Label	Previous	Current	Change (%)
2	09_Burglary (Residential)	624.6	646.1	3.4
2	10_Burglary (Other)	359.4	298.2	−17
2	11_Theft from Motor Vehicles	976.9	876.3	−10.3
2	12_Theft of Motor Vehicles	401.7	322.6	−19.7
2	13_Fatalities	6.3	5.9	−6.5
2	14_Serious Injuries	150.5	147.1	−2.2

Answer the following questions related to this data set.

12 What are the advantages of using percentages in tables (like those presented in Table 5.4) versus using raw numbers?

>

>

>

>

>

13 What are the disadvantages of using percentages?

>

>

>

>

>

14 If a news channel obtained a copy of the data in Table 5.4, how do you think it would use them?

>

>

>

>

>

15 Why do you think the 'Other offences' category has the biggest percentage increases in Table 5.4?

>

>

>

>

>

Answers can be found at the back of the book. Test your knowledge further by answering multiple choice questions at https://study.sagepub.com/jones

TIPS TO REMEMBER

 When constructing your own tables, make sure you think carefully about the headings used, as well as including units if needed. For example, if you are reporting the average weight from a sample of males in New York City, are you using kilograms or pounds?

 Also think about any legends or captions included in a table. It should include a description of what the table is displaying, the year the data was collected or constructed, and also the source (if it is a table you are using that someone else has made). An index or summary table could also highlight this information, as in the tables that have been used in this book under 'resources to support this chapter'.

 When reading tables, think about ways the data display could be improved. Would a graph be more appropriate? Could you use numbers or icons to display the data?

6

INTRODUCTION TO STATISTICS

Skills in this chapter

Data, data, data! It's everywhere you look! Data also comes in many different forms, which means it can be presented in a variety ways. Being able to identify

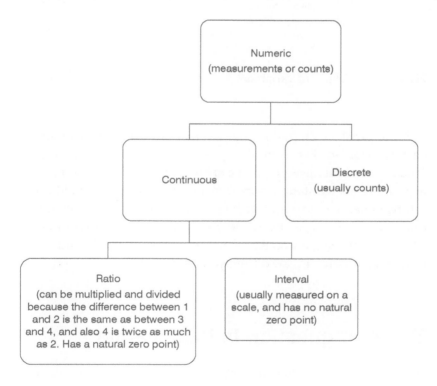

Figure 6.1 Types of variable 1

what type of data you have is extremely important, and one of the first steps in deciding what to do with it! Below is an overview of examples of variables, defined as any factor, trait, or condition that can exist in differing amounts or types. The chapters that follow this one will go into more detail about how to best present data, what questions to ask when looking at data, and other considerations to think about in terms of the data collection methods used.

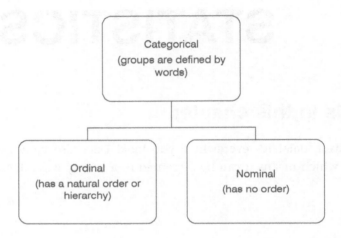

Figure 6.2   Types of variable 2

Some variables can fall into more than one group, depending on the way the data is recorded. For example, type of occupation can be a nominal variable (doctor, lawyer, shopkeeper) or it can be ordinal, having a natural order or hierarchy (e.g. shop floor assistant, assistant manager, manager). In addition, some variables can be presented differently to become other types of variables. For instance, age (normally a ratio type of variable) can be displayed as a set of ranges (e.g. 15–21, 22–25, 26–30), which would become a categorical variable, and it would be ordinal if the ranges were displayed from youngest to oldest.

Types of question in this chapter

Questions in this chapter will help to develop your skills in identifying the types of variables in the case studies presented.

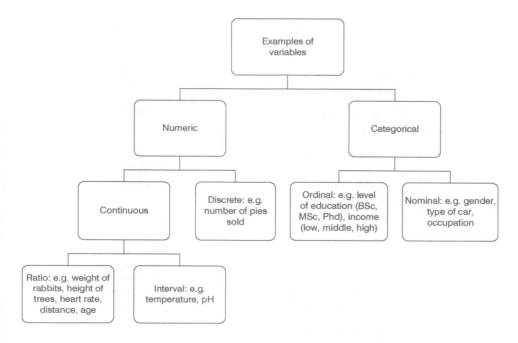

Figure 6.3 Types of variable 3

Resources to support this chapter

The data sets used in this chapter come from the Australian and South African government websites (see below). Table 6.1 contains Fire Service Brigade data collected from 2014, reduced to include fewer regions, from South Australia. Table 6.2 contains photography data from Ernest Gall, with works dating back to the late 1800s. The link to the original data set below also includes links to his photographic works, so take a look! Table 6.3 contains public health data on cancer rates in Australia, from 2007 to 2017. This data set has been shortened in size so it can be included in this chapter. Table 6.4 contains national energy data from South Africa in 2015. Reduction in size of the data set will enable you to develop your skills in reading tables, as well as being able to identify what types of data you are working with.

Case study	Content	Theme	Country and date downloaded
6.1	South Australian Country Fire Service Brigade incidents, 2018	Public services	Australia, 14 November 2018

https://search.data.gov.au/dataset/ds-sa-31a58078-8a02-43d5-b71a-c5c9cc47764f/details?q=

This data set comes from an Australian government website. Bush fires continue to be an environmental hazard to nations with vulnerable climates such as Australia. This data set lists the different hazards the fire services in South Australia had to deal with in 2018.

Case study	Content	Theme	Country and date downloaded
6.2	19th-century photographs by Ernest Gall, 1850–1910	Photography	Australia, 14 November 2018

https://search.data.gov.au/dataset/ds-sa-54c50844-6653-4704-8dc8-5f5c8f5f14eb/details?q=

This data set comes from an Australian government website. The data set includes information about the photographs that were taken by Ernest Gall, a South Australian-born photographer active from the 1880s to the 1920s. In 1899 he was described as a 'distinctly modern professional photographer'. This data set was chosen to show you different ways in which data can be presented.

Case study	Content	Theme	Country and date downloaded
6.3	South Australian Cancer Registry, 2007–2017	Public health	Australia, 14 November 2018

https://search.data.gov.au/dataset/ds-sa-60aca0c6-0c84-4bac-8359-d736a-d4a67e7/details?q=

This data set comes from an Australian government website. Cancer continues to be a big cause of mortality and morbidity worldwide. Treatment and detection methods have improved dramatically, which requires healthcare professionals to constantly monitor cancer rates and disease progression, over time. This data set presents cancer rates and deaths over a period of time in South Australia.

Case study	Content	Theme	Country and date downloaded
6.4	South African annual aggregated fuel sales, 2015	Energy	South Africa, 14 November 2018

http://www.energy.gov.za/files/media/media_SAVolumes.html

This data set comes from a South African government website. Climate change and energy consumption are hot topics for governments and businesses to consider. This data set looks at the fuel sales in South Africa in 2015.

COMMON ERRORS

 Be careful when reading from large data sets. It is very easy to mix up values, which could lead to incorrect calculations.

 Pay close attention to the types of data in each of the tables, arranged in columns and rows.

The climate in Australia is notorious for reaching extremely high temperatures, making bush fires especially hazardous for nearby inhabitants. Fire services are an essential asset in being able to deal with problems caused by adverse weather conditions, and other situations that can lead to danger to humans and animals. The data set in Table 6.1 presents a series of events that required Fire service intervention.

Table 6.1 South Australian Country Fire Service Brigade incidents, 2018

Brigade	Region	Situation found
Cudlee Creek	2	Severe weather and natural disaster
Happy Valley	1	Building fire – content only
Mount Barker	1	Vehicle accident / no injury
Hahndorf	1	Cooking fumes (toast or foodstuffs) – 756
Aldinga Beach	1	Alarm sounded no evidence of fire
Roseworthy	2	Did not arrive (stop call)
Ardrossan	2	Heat/thermal detector operated, no fire/heat from source – 752
Mount Barker	1	Arcing, shorted electrical equipment
Dublin	2	Rubbish, refuse or waste – abandoned outside
Coomandook	3	Mobile property/vehicle
Gumeracha	2	Tree down
Upper Sturt	1	Tree down
Murray Bridge	3	FIP – reset prior to arrival by management – 730
McLaren Vale	1	Alarm sounded – no evidence of fire
Angaston	2	Alarm system suspected malfunction – 739
Goolwa	1	Extrication/rescue (not vehicle)
Aldinga Beach	1	Alarm activation by outside tradesman/occupier activities – 765
Burra	4	Cooking fumes (toast or foodstuffs) – 756
Stirling	1	Vehicle accident rescue
Strathalbyn	1	Building fire
Mount Barker	1	Building fire
Woodside	1	Cooking fumes (toast or foodstuffs) – 756

Brigade	Region	Situation found
Gawler River	2	Mobile property/vehicle
Virginia	2	Mobile property/vehicle
Inman Valley	1	Tree down
Lochiel	4	Vehicle accident/no injury
Aldinga Beach	1	Building fire – structure and content
Echunga	1	Building fire
Millicent	5	Hazardous material
Eden Hills	1	Vehicle accident with injuries

Answer the following questions related to this data set.

1 What type of variable is Situation found?

>

>

>

>

>

2 Do you think the Situation found variable has enough information for Fire services staff to adequately deal with the incident?

>

>

>

>

>

3 What additional information, apart from the variables in Table 6.1, do you think would be helpful for Fire services staff to deal with the Situation found?

>

>

>

>

>

4 What type of variable is Region?

>

>

>

>

>

5 Which Region in Table 6.1 appears to have the most incidents reported?

>

>

>

>

>

Answers can be found at the back of the book. Test your knowledge further by answering multiple choice questions at https://study.sagepub.com/jones

Photographers work to capture a moment or event that elicits certain emotions in us when we look at their works. Landscape pictures are able to capture areas of beauty, often resonating strongly with past memories. This data set presents information about a series of photographs taken by Ernest Gall, a South Australian photographer active from the 1880s to the 1920s.

Table 6.2   19th century photographs by Ernest Gall, 2017

Title	Dates/Publication details	Description/Quantity	Series/Collection
Railway Station, Blackwood	[approximately 1900]	Photograph 15 cm × 10.5 cm	Blackwood Collection
St James Church, Blakiston	1895	Photograph & neg. 15.6 cm × 10.7 cm	Blakiston Collection
Institution, Brighton	1889	Photograph 15.2 cm × 10.8 cm	Brighton Collection
Burra	1850	Photograph & neg. 14.4 cm × 7.8 cm	Burra Collection
Clarendon	[approximately 1890]	Photograph 15 cm × 10.7 cm	Clarendon Collection
River, Clarendon	[approximately 1890]	Photograph 15 cm × 10.7 cm	Clarendon Collection
Bridge, Clarendon	1889	Photograph 15 cm × 10.7 cm	Clarendon Collection
Clarendon	[approximately 1889]	Photograph 15 cm × 10.7 cm	Clarendon Collection
Cottage, Clarendon	1889	Photograph 15 cm × 10.7 cm	Clarendon Collection
Royal Oak Hotel, Clarendon	1910	Photograph & neg. 21.1 cm × 15.5 cm	Clarendon Collection
Coromandel Valley	[approximately 1889]	Photograph 19.8 cm × 14.3 cm; Reserve 1 b	Coromandel Valley Collection
Coromandel Valley	1889	Photograph 15 cm × 10.7 cm	Coromandel Valley Collection

(Continued)

Table 6.2 (Continued)

Title	Dates/Publication details	Description/Quantity	Series/Collection
Institute, Coromandel Valley	1889	Photograph 15 cm × 10.7 cm	Coromandel Valley Collection
Enfield	[approximately 1861]	photograph	Enfield Collection
Gilles Plains	1906	Photograph & neg. 24.4 cm × 19.5 cm	Gilles Plains Collection
Jetty Road, Glenelg	[approximately 1904]	Photograph 10.0 cm × 6.3 cm	Glenelg Collection
Glenelg	1896	Photograph & neg. 20.7 cm × 15.3 cm	Glenelg Collection
Glenelg	1896	Photograph & neg. 21.6 cm × 15.3 cm	Glenelg Collection
Glenelg	1896	Photograph & neg. 20.5 cm × 15.3 cm	Glenelg Collection
Glenelg	1904	Photograph 10.0 cm × 7.4 cm	Glenelg Collection
Glenelg	1904	Photograph 10.0 cm × 7.4 cm	Glenelg Collection
Glenelg	[approximately 1896]	Photograph (B&W print) 14.7 cm × 20 cm; Reserve 1	Glenelg Collection
Glen Osmond	1889	Photograph 15 cm × 10.7 cm	Glen Osmond Collection
St. Peter's College, Hackney	1889	Photograph 14.7 cm × 10.6 cm	Hackney Collection

Answer the following questions related to this data set.

6 The Description/Quantity variable includes two pieces of information. What type of variable is Quantity?

>

>

>

>

>

7   If you multiply the two length measurements together assuming that they represent width and length, what value would this give you?

>

>

>

>

>

8   What type of variable is Title?

>

>

>

>

>

9   Which Title has the most entries in this data set?

>

>

>

>

>

10   Could you improve the clarity of Table 6.2? If yes, in what ways?

>

>

>

>

>

Answers can be found at the back of the book. Test your knowledge further by answering multiple choice questions at https://study.sagepub.com/jones

Case study 6.3 presents data on all South Australian residents from 2007 to 2017. Data includes number of New cases of and Deaths caused by cancer during this time.

Table 6.3   South Australian Cancer Registry, 2007–17

Year	2007	2008	2009	2010	2011	2012	2013	2014	2015	2016	2017
New cases	9,079	9,366	9,297	9,529	9,398	9,486	9,717	10,073	10,353	10,512	10,695
Deaths	3,508	3,685	3,533	3,657	3,486	3,549	3,571	3,629	3,662	3,654	3,674

Answer the following questions related to this data set.

11   What type of variable is New cases?

>

>

>

>

>

12   What type of variable is Deaths?

>

>

>

>

>

13   Which year in Table 6.3 had the highest number of Deaths?

>

>

>

>

>

14 Which year had the lowest number of New cases?

>

>

>

>

>

15 What type of variable is Year?

>

>

>

>

>

Answers can be found at the back of the book. Test your knowledge further by answering multiple choice questions at https://study.sagepub.com/jones

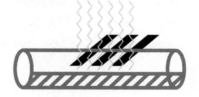

The following data set presents data on the total volume of national fuel sales for South Africa in 2015. The fuel sales are grouped into different types of fuel, and include total values for all fuel types. Fuel costs can have huge knock-on effects to other services and products offered to consumers; for example, food and drink prices are partly dependent on shipping and transportation costs.

Table 6.4   South African annual fuel sales, 2015

		Volume in litres			
Product name	**January– March (Q1)**	**April– June (Q2)**	**July– September (Q3)**	**October– December (Q4)**	**Grand total**
Diesel (all grades)	3,628,347,660				3,628,347,660
Petrol (all grades)	3,048,003,166				3,048,003,166
Jet fuel	613,998,590				613,998,590
Paraffin	142,867,450				142,867,450
Furnace oil	149,268,822				149,268,822
LPG	136,629,850				136,629,850
Aviation gasoline	4,334,718				4,334,718
Grand total	7,723,450,256				7,723,450,256

Answer the following questions related to this data set.

16   What type of variable is Volume in litres?

>

>

>

>

>

17   Which Product name had the highest Volume in litres in the first quarter?

>

>

>

>

>

18 Express the highest Volume in litres value as a proportion of the Grand total.

>

>

>

>

>

19 What type of variable is Product name?

>

>

>

>

>

20 What is the difference between the highest and smallest Volume in litres values in Table 6.4?

>

>

>

>

>

Answers can be found at the back of the book. Test your knowledge further by answering multiple choice questions at https://study.sagepub.com/jones

Think about how each data set could be made clearer by being cleaned up and presented in a different way. What pieces of information are essential, and what parts of the table can be cut out?

Make sure you are able to identify what type of data you have, and that you are able to label it as being either categorical or numeric.

7
GRAPHS

Skills in this chapter

Using graphs is a very common way of presenting data, found in many disciplines. In this chapter, you will learn how to interpret graphs, and we will cover the main types used in statistics. Being able to produce graphs from data usually enables the user to spot features in the data much more easily. Whenever you are looking at a graph, you need to ask yourself these questions:

- What are the main features I can see in the graph?
- What does this mean?
- What other details are useful for understanding the variable or answering any questions that I may have?
- What other questions do I have?

You may not be able to answer all these questions; however, they are useful ways to help you to think about data critically.

The type of graph used to display data will depend on what type of data you have. For numeric data, useful graphs to display this information can include histograms, scatter plots, dot plots, box-and-whisker plots. If you have categorical data, bar graphs and side-by-side plots are often useful.

The best type of graph to illustrate the relevant data will also depend on the numbers of variables you are comparing, as well as the number of groups. This will be covered in more detail in the next chapter.

Much of the data we have covered in this book has included the measurement of variables (numeric or categorical) over time, which we call time series data. This is especially relevant for disciplines that investigate changes over time, such as the sciences, social sciences, psychology, history and public health. Time series data can include changes over short or long periods. This chapter will look at several case studies that examine changes in variables over time.

The main types of graph used to display data are shown below, including guidance on what the respective parts of the graph are telling you.

Histograms

Figure 7.1 shows a typical histogram.

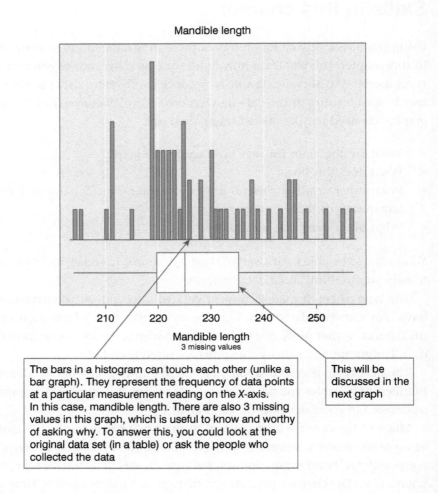

The bars in a histogram can touch each other (unlike a bar graph). They represent the frequency of data points at a particular measurement reading on the X-axis. In this case, mandible length. There are also 3 missing values in this graph, which is useful to know and worthy of asking why. To answer this, you could look at the original data set (in a table) or ask the people who collected the data

This will be discussed in the next graph

Figure 7.1 Histogram

Dot plots, box-and-whisker plots (sometimes just called box plots) and scatter plots

Figures 7.2–7.4 are based on the dog walking areas data in Table 3.3.

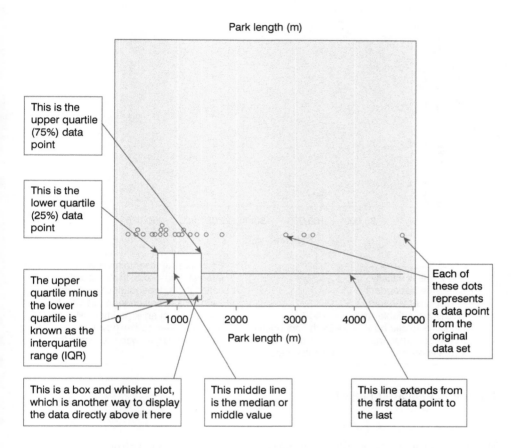

Park length (m)

This is the upper quartile (75%) data point

This is the lower quartile (25%) data point

The upper quartile minus the lower quartile is known as the interquartile range (IQR)

Each of these dots represents a data point from the original data set

This is a box and whisker plot, which is another way to display the data directly above it here

This middle line is the median or middle value

This line extends from the first data point to the last

0   1000   2000   3000   4000   5000

Park length (m)

Figure 7.2 Dot plots and box-and-whisker plots

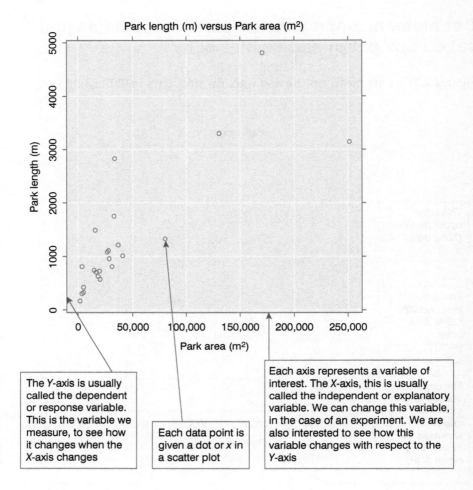

Figure 7.3 Scatter plots 1

When looking at scatter plots, we need to think about four things:

Trend. Is there a linear (straight-line) relationship – in other words, can you fit roughly all the dots onto a straight line? On the other hand, do the dots appear to fit a curved line? In which case we could say there appears to be a nonlinear relationship. Alternatively, if the dots do not seem to show any pattern, we could say there is no relationship.

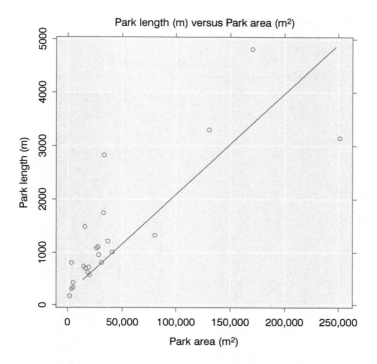

Figure 7.4 Scatter plots 2

Scatter. Are the data points constant? If you were to draw a straight line though the data points, are they close to the line? Alternatively, is there a lot of variation in data points around the line? For example, look at the scatter plot with the blue line drawn on top of the data points (Figure 7.4). If the scatter were constant in this graph, we would expect the data points to be closer to the line, and not so spread out.

Strength of relationship. If the data points are close to a line we draw as closely as possible to all the data points that we place on top, then we can say there is a strong relationship between the variables on the axis. If the data points are spread out and far away from each other, with no discernible pattern, then we say there is a weak relationship.

Association. If the data points on the *Y*-axis increase in line with the *X*-axis, as is the case in Figure 7.4, then we can say there is a positive association. If, however, the data points decrease on the *Y*-axis as the *X*-axis increases, then we say there is a negative association.

Bar graphs and side-by-side plots

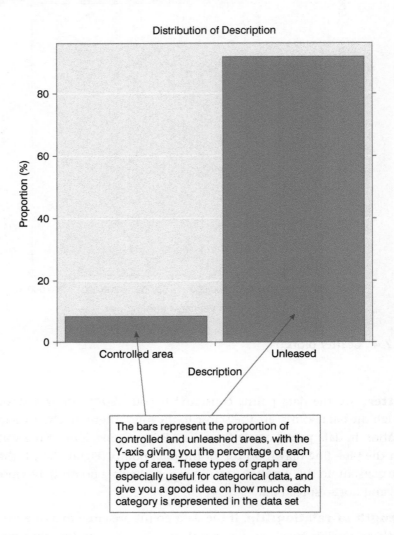

Figure 7.5 Bar graphs

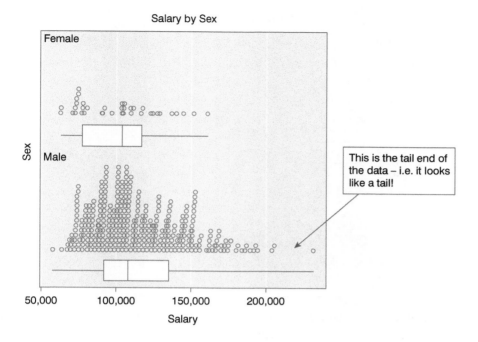

Figure 7.6 Side by side dot plots and box and whisker plots – on the same scale

Side-by-side plots are particularly useful when wanting to compare a numerical value, or data from different categories (in Figure 7.6, males and females).

Other questions you should ask yourself when looking at graphs will depend on the type of graph you are looking at, to be able to spot a particular feature about the data. The following features are a useful starting point:

Skewness. When looking at dot plots or histograms, how the data is shaped will tell you if it is positively skewed, negatively skewed or symmetrical. For example, if we look at the graph that splits up salary by sex (side-by-side plots) the tail end of the male data is located towards the positive X-axis values (Figure 7.7). We therefore call this positive skew. If the tail end of the male data points were located towards the increasingly negative X-axis part of the graph, we would call this negative skew.

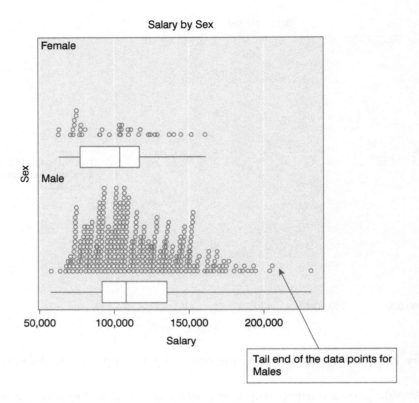

Figure 7.7 Side by side dot plots and box and whisker plots – on the same scale

Symmetry. Assessing the symmetry of your data points, which is best performed on dot plots or histograms, can be thought of as placing a line somewhere in the middle of the data points, and seeing if it mirrors each side. For example, if you could place a line in the middle of your data points on a piece of paper, and you folded it along the line you have drawn, would they roughly fit over each other?

Modality. The modality of data refers to the most frequent points in your data set. For the graph of male and female salaries, the male graph looks roughly unimodal – in other words, there is one point on the graph of male salaries that is more frequent than the rest. If there were two clear modal values, and the data spread looked more like a camel's back with two humps, we would say the data has a bimodal distribution. We will meet modality again in Chapter 8.

Outliers. An outlier is a data point that appears to be far away from the rest of your data points. There is sometimes the temptation to remove outliers from a data set, which should be avoided! It is much better to ask questions about that data point. Why is it so far away from the others? Could there have been a mistake in reading the data? On the other hand, is the outlier of interest? Removing a perceived outlier can have big implications for any statistical analysis you perform on the data set, and might not give you the true answers.

Table 7.1 provides a useful overview of the strengths and weaknesses of commonly used graphs.

Table 7.1   Strengths and weaknesses of commonly used graphs

Type	Strengths and weaknesses	Example
Dot plot	• Retains numerical information, can also check for skewness of data, modality and outliers	Gender and salary
Scatter plot	• Can give you useful information on whether the axes in the graph are associated with each other • Retains numerical information	Age and salary
Box-and-whisker plot	• Very good for comparing several data sets • Displays centre and spread (see Chapter 8) • Not useful for small data sets	Gender and salary
Histogram	• Displays relative density of observations (i.e. gives a good idea of the • shape of the distribution) • Good for large amounts of data	Heights

Types of question in this chapter

The questions in this chapter will build on developing your skills in identifying the types of variables present in the case studies presented. Taking this a step further, we will introduce you to techniques in visualising data using the most suitable graphs. We will also develop your abilities in interpreting graphs.

Resources to support this chapter

The data sets used in this chapter come from the Japanese and Australian government websites (see below). Table 7.2 contains population data collected from 2012 to 2014, and Table 7.3 contains waste data collected from 2005 to 2013, both translated from Japanese into English (using Google Translate) and reduced to include fewer data groups, from the city of Etajima in Hiroshima prefecture. Table 7.4 contains crime data for the region of South Australia, in 2018. This data set, shortened from the original set, was first introduced in Chapter 5. The graph generated based on this data uses the complete data set, which includes six regions.

Case study	Content	Theme	Country and date downloaded
7.1	Waste produced by Etajima city, Hiroshima prefecture – 2005–2013	Environment	Japan, 23 March 2018

http://www.city.etajima.hiroshima.jp/cms/articles/show/4457

This data set comes from a Japanese government website. Waste from households and businesses and their disposal are becoming increasingly difficult to deal with by governments. This data set looks at the different types of waste in Etajima city, and changes in the amount of these waste types, over time.

Case study	Content	Theme	Country and date downloaded
7.2	Population of Etajima city, Hiroshima prefecture – 2012–2014	Population studies	Japan, 23 March 2018

http://www.city.etajima.hiroshima.jp/cms/articles/show/4457

This data set comes from a Japanese government website. The data set was chosen to present the reader with a source of information that can help inform may high level decisions for policy makers, government and also businesses. Changes to population levels within countries can have a profound effect on resource allocation for example.

Case study	Content	Theme	Country and date downloaded
7.3	South Australian Country Fire Service Brigade incidents, 2018	Public services	Australia, 14 November 2018

https://search.data.gov.au/dataset/ds-sa-31a58078-8a02-43d5-b71a-c5c9cc47764f/details?q=

This data set comes from an Australian government website. Bush fires continue to be an environmental hazard to nations with vulnerable climates such as Australia. This data set lists the different hazards the fire services in South Australia had to deal with in 2018.

COMMON ERRORS

 Pay close attention to the information given in the graph. Look at the axes, see what they represent, and also look at the units.

 When selecting an appropriate graph to display your data, ensure you have identified what type of data you have first. This will enable you to select the most appropriate type of graph to use.

 Think about the scale that you are using to display your data, in the graph you have selected. Using a scale that is too big or too small could mean the message you are trying to convey in your data gets lost.

The data presented in Table 7.2, and Figures 7.8–7.10, show the different types of waste produced by Etajima city, Hiroshima prefecture, Japan, from 2005 to 2013. The data has been translated from the original data file, which was in Japanese. This type of data could be extremely useful for government departments responsible for waste collection and maintaining recycling plants. Being able to assess changes in each type of waste over time can help to assess whether waste collection for the region is frequent enough, or whether there needs to be a change in the way waste is collected. For example, recycling centres at supermarkets could encourage people to bring their waste with them, reducing the need for it to be collected from their homes.

Volume of waste in tonnes

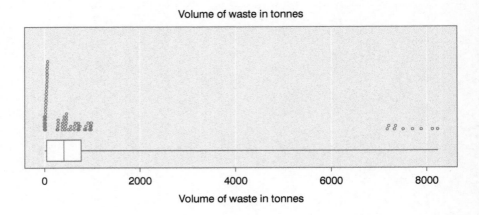

Volume of waste in tonnes

Figure 7.8 Dot plot and box and whisker plot of Volume of waste (in tonnes), Etajima city, Hiroshima prefecture, Japan, 2005–2013

Answer the following questions related to this data.

1   What is the median value in the graph of Volume of waste?

>

>

>

>

>

Table 7.2 Waste (in tonnes) produced by Etajima city, Hiroshima prefecture, Japan, 2005–2013

Waste type	2005	2006	2007	2008	2009	2010	2011	2012	2013
Burnable	8,227.7	8,127.25	7,889.85	7,710.99	7,511.84	7,175.45	7,349.94	7,206.22	7,326.48
Paper	953.03	973.18	913.04	868.9	958.26	983.983	975.11	841.38	741.47
Cloth	0	0	0	0	0	31.06	44.5	38.36	35.24
Incombustible	403.05	372.39	265.97	279.15	282.24	287.49	287.19	384.36	376.01
Bulky	701.69	628.26	548.636	711.92	726.92	869.67	707.03	626.66	651.73
Recyclable garbage	553.441	506.88	462.504	457.39	463.82	458.42	442.09	426.82	413.19
PET bottles	55.48	66.25	58.1	54.05	55.09	55.49	52.65	47.42	46.84
Metal	1,685.18	2,080.98	1,351.97	1,146.76	5,363.8	1,198.72	698.16	1,325.01	613.4
Other harmful waste	24.453	21.72	28.156	19.43	19.47	20.45	19.89	19.722	19.53
Total	12,604.02	12,776.91	11,518.22	11,248.59	15,381.44	11,080.73	10,576.56	10,915.95	10,223.89

2 Are there any outliers in the dot plot?

>

>

>

>

>

3 Would you say the data in the dot plot are skewed?

>

>

>

>

>

4 Would you say the data points in the dot plot are uniformly distributed?

>

>

>

>

>

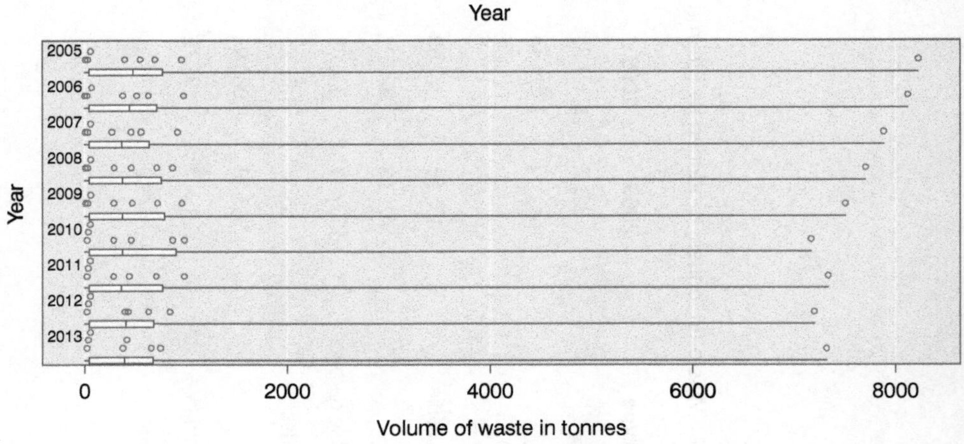

Figure 7.9 Side by side dot plot and box and whisker plot of Volume of waste (in tonnes), Etajima city, Hiroshima prefecture, Japan, 2005–2013

5 Which year shows the most variation in data spread?

>

>

>

>

>

6 Which year shows the greatest difference between the lower and upper quartiles?

>

>

>

>

>

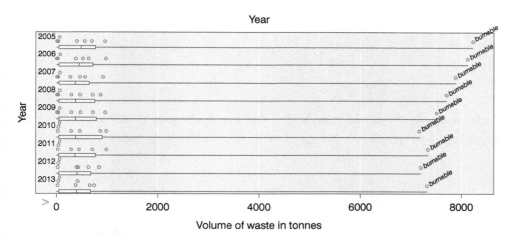

Figure 7.10 Side by side dot plot and box and whisker plot of Volume of waste (in tonnes), Etajima city, Hiroshima prefecture, Japan, 2005–2013

7 Of all the years of data collection, which year had the highest Volume of waste in tonnes?

>

>

>

>

>

8 For burnable waste, would you say the levels from 2005 to 2013 are increasing or decreasing?

>

>

>

>

>

9 Did you have trouble in using the dot plots above? If you did, what sort of trouble?

>

>

>

>

>

Answers can be found at the back of the book. Test your knowledge further by answering multiple choice questions at https://study.sagepub.com/jones

The data presented in Table 7.3 and Figure 7.11 shows the population of Etajima city, Hiroshima prefecture, Japan, in 2013 and 2014. Monitoring changes in population composition (i.e. proportions of males to females, as well as actual changes in numbers of males and females) will have all sorts of implications for the region in which these people reside. For example, are there enough people being born in a region to ensure the population is being maintained?

Table 7.3 Population of Etajima city, Hiroshima prefecture, Japan, 2013 and 2014

Region	2013				2014			
	Male	Female	Total	Number of households	Male	Female	Total	Number of households
1	687	840	1,527	731	664	818	1,482	718
2	528	445	973	524	476	434	910	505
3	310	315	625	288	316	298	614	281
4	244	305	549	284	234	296	530	278
5	700	833	1,533	768	682	808	1,490	760
6	884	997	1,881	891	867	986	1,853	883
7	36	43	79	36	32	39	71	35
8	103	137	240	118	105	126	231	118
9	313	308	621	264	318	302	620	281
10	399	505	904	447	386	488	874	432
11	184	20	204	204	185	14	199	198
12	982	1,079	2,061	953	1,002	1,084	2,086	964
13	559	673	1,232	593	554	661	1,215	586
14	231	280	511	279	225	276	501	272
15	263	297	560	297	258	288	546	292
16	532	598	1,130	544	517	590	1,107	537
17	243	256	499	219	234	251	485	212
18	160	182	342	180	156	180	336	180
19	366	437	803	414	352	426	778	405

(Continued)

Table 7.3 (Continued)

Region	2013				2014			
	Male	Female	Total	Number of households	Male	Female	Total	Number of households
20	1,133	1,229	2,362	1,092	1,119	1,232	2,351	1,092
21	426	473	899	455	410	457	867	442
22	792	819	1,611	794	766	778	1,544	774
23	661	733	1,394	668	680	746	1,426	685

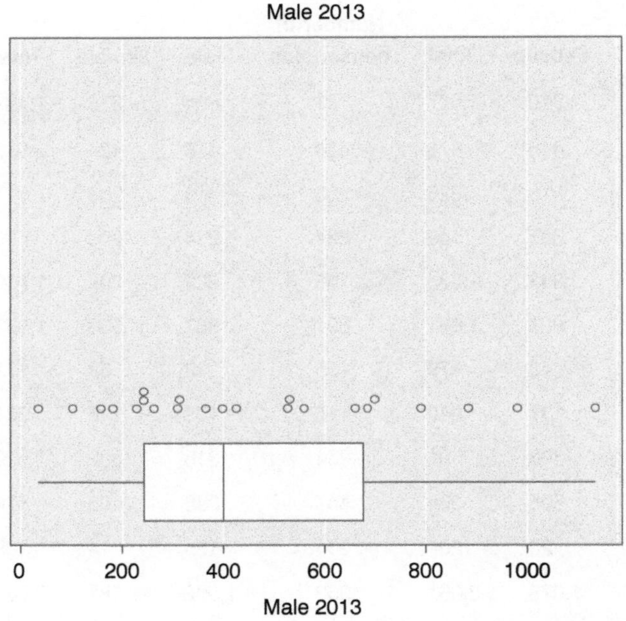
Male 2013
0
200
400
600
800
1000
Male 2013

Figure 7.11 Dot plot and box and whisker plot of the population of Etajima city, Hiroshima prefecture, Japan, 2012–2014.

Answer the following questions related to this data.

10 What is the median value for Males in 2013?

>

>

>

>

>

11 What is the highest value of Males in 2013?

>

>

>

>

>

12 What are the advantages of using a dot plot and box-and-whisker plot, over a table?

>

>

>

>

>

Answers can be found at the back of the book. Test your knowledge further by answering multiple choice questions at https://study.sagepub.com/jones

The climate in Australia is notorious for reaching extremely high temperatures, making bush fires especially hazardous for nearby inhabitants. The fire services are an essential asset in being able to deal with problems caused by adverse weather conditions, and other situations that can lead to danger to humans and animals. The data set in Table 7.4 presents a series of events that required Fire Service intervention, followed by a summary statistics table (Figure 7.12), outlining the different regions present in Table 7.4, and the percentage of situation found items that appear in each region. Finally, the bar graph in Figure 7.13 illustrates this data. The data displayed in Figures 7.12 and 7.13 come from the complete data set (i.e. it includes six regions), whereas the data in Table 7.4 is a shorter version of this complete data set.

Table 7.4   South Australian Country Fire Service Brigade incidents, 2018

Brigade	Region	Situation found
CUDLEE CREEK	2	SEVERE WEATHER AND NATURAL DISASTER
HAPPY VALLEY	1	BUILDING FIRE – CONTENT ONLY
MOUNT BARKER	1	VEHICLE ACCIDENT/NO INJURY
HAHNDORF	1	COOKING FUMES (TOAST OR FOODSTUFFS) – 756
ALDINGA BEACH	1	ALARM SOUNDED NO EVIDENCE OF FIRE
ROSEWORTHY	2	DID NOT ARRIVE (STOP CALL)
ARDROSSAN	2	HEAT/THERMAL DETECTOR OPERATED, NO FIRE, HEAT FROM SOURCE – 752
MOUNT BARKER	1	ARCING, SHORTED ELECTRICAL EQUIPMENT
DUBLIN	2	RUBBISH, REFUSE OR WASTE – ABANDONED OUTSIDE
COOMANDOOK	3	MOBILE PROPERTY/VEHICLE
GUMERACHA	2	TREE DOWN
UPPER STURT	1	TREE DOWN
MURRAY BRIDGE	3	FIP – RESET PRIOR TO ARRIVAL BY MANAGEMENT – 730
MCLAREN VALE	1	ALARM SOUNDED – NO EVIDENCE OF FIRE
ANGASTON	2	ALARM SYSTEM SUSPECTED MALFUNCTION – 739
GOOLWA	1	EXTRICATION/RESCUE (NOT VEHICLE)
ALDINGA BEACH	1	ALARM ACTIVATION BY OUTSIDE TRADESMAN/OCCUPIER ACTIVITIES – 765

Brigade	Region	Situation found
BURRA	4	COOKING FUMES (TOAST OR FOODSTUFFS) – 756
STIRLING	1	VEHICLE ACCIDENT RESCUE
STRATHALBYN	1	BUILDING FIRE
MOUNT BARKER	1	BUILDING FIRE
WOODSIDE	1	COOKING FUMES (TOAST OR FOODSTUFFS) – 756
GAWLER RIVER	2	MOBILE PROPERTY/VEHICLE
VIRGINIA	2	MOBILE PROPERTY/VEHICLE
INMAN VALLEY	1	TREE DOWN
LOCHIEL	4	VEHICLE ACCIDENT/NO INJURY
ALDINGA BEACH	1	BUILDING FIRE – STRUCTURE AND CONTENT
ECHUNGA	1	BUILDING FIRE
MILLICENT	5	HAZARDOUS MATERIAL
EDEN HILLS	1	VEHICLE ACCIDENT WITH INJURIES

Summary

Primary variable of interest: Region (categorical)

Total number of observations: 29824

Summary of the distribution of Region:

	1 – REGION 1	2 – REGION 2	3 – REGION 3	4 – REGION 4	5 – REGION 5	6 – REGION 6	Total
Count	11570	9220	2947	2489	2270	1328	29824
Percent	38.794%	30.915%	9.881%	8.346%	7.611%	4.453%	100%

Figure 7.12 Summary statistics diagram

CASE STUDY 7.3

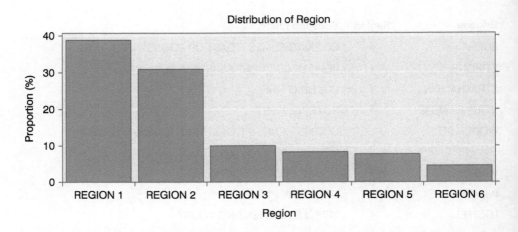

Figure 7.13 Bar graph of Region

Answer the following questions related to this data.

13 Which Region had the highest percentage of Situation found items?

>
>
>
>
>

14 Which Region had the lowest percentage of Situation found items?

>
>
>
>
>

15 Could there be a reason why Region 1 had more Situation found items than Region 6?

>
>
>

>

>

16 How would you display the Situation found variable presented in Table 7.4?

>

>

>

>

>

17 What problems might you face when trying to visualise the Situation found variable?

>

>

>

>

>

18 Are there any other types of graph you have come across, not mentioned in this chapter? If yes, what are they?

>

>

>

>

>

19 Why do you think we use different types of graph for different types of data?

>

>

>

>

>

20 Do you think it is important to think about the type of data that you want to collect, before you think of a method for collecting it? Explain your answer.

>

>

>

>

>

Answers can be found at the back of the book. Test your knowledge further by answering multiple choice questions at https://study.sagepub.com/jones

TIPS TO REMEMBER

Whenever you are looking at data, think about it critically. How can you visualise this data to help you spot patterns? Remember the essential questions you should ask:

What are the main features I can see in the graph?

What does this mean?

What other details are useful for understanding the variable or answering any questions that I may have?

What other questions do I have?

8

EXPLORING DATA AND DESCRIPTIVE STATISTICS

Skills in this chapter

We're at a point in our data journey where it's time to bring together several of the skills you have developed, reinforcing the work you have completed on percentages, proportions, reading tables and graphs. This chapter will take you a step further, getting you to think about how data is distributed, as well as comparing different types of averages. We will also get you to think about data critically, weighing up the pros and cons of using different ways to present and interpret data. Much of what we will cover in this chapter comes under the umbrella term 'descriptive statistics'. Descriptive statistics are used to describe the basic features of data in a data set. They provide simple summaries about the sample or population. With descriptive statistics, you are simply describing what the data is or what it shows.

Data organisation

Data is often organised as a rectangular data set. In a rectangular data set, each row corresponds to an entity and each column corresponds to a variable. Throughout this book, you have come across these types of data sets in all the tables you have worked through.

The entities we collect and store in individual-level data could be anything: people, cities, countries, animals, plants, companies, financial transactions, telephone calls, and so on.

Variables are the properties being recorded about each entity. We have looked at many variables throughout this book, such as Park length, Situation found, Age group and Region.

Manipulating data sets to transform them into a format so they can be processed by data analysis software can take a considerable amount of time. Often two or more different data sets need to be merged together to create a more complex set of data to explore. These skills are important, but not discussed in this book. However, you need to consider whether in the process of manipulating data sets, these actions taken could have affected the quality of the data.

Exploring data

The exploration of data enables us to pose and even answer interesting questions about the world around us. Data is everywhere, present in all disciplines, and can be manipulated, transformed and reported in many ways. You have already looked at ways to read and report data in tables, and you have looked at the advantages of producing graphs from data in tables. This chapter will bring all these skills together and get you to think about how data in different forms can give different perspectives on how to interpret the data presented. This chapter will build upon your critical thinking skills, helping you to evaluate the pros and cons of using different approaches to present and interpret data.

Remember the questions we asked ourselves in Chapter 7: these are extremely useful questions to help us to interrogate data, and we can use them to question data in many forms (tables as well as graphs, for example).

....

- What are the main features I can see in the graph or table?

- What does this mean?

- What other details are useful for understanding the variable or answering any questions that I may have?

- What other questions do I have?

....
....

Descriptive statistics

Descriptive statistics help us to get an idea of the central tendency of a data set. Central tendency is a fancy way of saying where the middle of the data sits. Table 8.1 lists three ways to measure it.

Table 8.1 Central tendency

Centre	Summary statistics	Notes
Where the 'middle' of the set of observations is	**Median** This is the middle value when the values are arranged in order (covered in Chapter 7)	• If we had to summarise all the observations on a variable as a single number, then we'd want to use a measure of their 'centre'
The two most commonly used measures of centre are the *median* and the *mean*	**Mean** This is a statistical name for the ordinary, everyday average. It is where the dot plot balances **Mode** The most frequently occurring data point in the set of observations	• If the shape is roughly symmetric, the mean and the median will be approximately the same • If the shape is strongly skewed there is no single compelling notion of 'centre' and the mean and median can be quite different • Means can be skewed by outliers in smaller data sets whereas medians are not

Table 8.2 Spread

Spread	Summary statistics
How 'spread out' the observations are along the scale, i.e. how variable the observations are, or how different they are from one another	**Standard deviation** This can be thought of as 'the average of the distances between the points and the mean'
Smaller spread: the observations are less variable, i.e. less different from one another	When a dot plot is a roughly symmetric mountain shape then typically *about 68% of the data points fall within one standard deviation either side of the mean*
Larger spread: the observations are more variable, i.e. more different from one another Commonly used measures of the spread are the *interquartile range* (IQR) (covered in Chapter 7) and the *standard deviation*	In detail, the standard deviation is 'the square root of the average of the squared distances between the points and their mean' (average the squared distances and then take the square root of the answer)

But the central tendency isn't the whole story. We are often interested in how spread out the data are. If we put the data into a graph, what does the data spread look like? Is it symmetrical?

Calculating a mean and standard deviation for a sample or population can be done by hand. However, there are many online calculators that will do this for you – for example, https://www.calculator.net/standard-deviation-calculator.html.

When using standard deviation values, you need to check whether the data set you are using is a sample from a larger population, or whether it is from the whole population. The standard deviation of a population and a sample are usually different.

Types of question in this chapter

Questions in this chapter will build on developing your skills in reviewing data, thinking about data presentation and interpretation. These questions will help to reinforce skills you have developed based on percentages, proportions, tables and graphs. Many of the questions will also draw on the context for each case study, getting you to think about what the data means in the real world. For example, are there cultural or political issues to consider?

Resources to support this chapter

The data sets used in this chapter come from the government website of Hong Kong (see below). All the tables used in this chapter come from one source, which looks at different industry sectors in Hong Kong, across different periods. The data sets have been shortened, to help you focus in on fewer variables and encourage you to develop your skills in critical thinking and data handling.

Case study	Content	Theme	Country and date downloaded
8.1	Industry sections and number of persons engaged in the Sai Kung district, Hong Kong 2018	Labour market and population studies	Hong Kong, 29 January 2019

Case study	Content	Theme	Country and date downloaded

https://data.gov.hk/en-data/dataset/hk-censtatd-tablechart-eandc

This data set comes from a Hong Kong government website. Monitoring changes in the labour market can provide insightful data for businesses and governments. In particular, it can help to identify trends over time, and could help to inform marketing campaigns to increase recruitment in certain industry areas.

Case study	Content	Theme	Country and date downloaded
8.2	District, number of establishments and number of persons engaged in education, Hong Kong 2000	Labour market and population studies	Hong Kong, 29 January 2019

https://data.gov.hk/en-data/dataset/hk-censtatd-tablechart-eandc

This data set comes from a Hong Kong government website. Hong Kong ranks highly in international league tables, for the quality of its education at all levels, and the achievement of its students. This data set looks at the number of educational establishments in Hong Kong, along with the number of people working at these places.

Case study	Content	Theme	Country and date downloaded
8.3	District, number of establishments and number of persons engaged in the financing and insurance industry, Hong Kong, 2008 and 2009	Labour market and population studies	Hong Kong, 29 January 2019

https://data.gov.hk/en-data/dataset/hk-censtatd-tablechart-eandc?q=

This data set comes from a Hong Kong government website. Hong Kong is one of the central financial hubs on the planet. Data based on finance related establishments, along with the number of people working in these places, in presented in this data set.

(Continued)

Case study	Content	Theme	Country and date downloaded
8.4	Percentage difference of persons engaged in the financing and insurance industry in Hong Kong, 2008 to 2009	Labour market and population studies	Hong Kong, 29 January 2019

https://data.gov.hk/en-data/dataset/hk-censtatd-tablechart-eandc

This data set comes from a Hong Kong government website. Hong Kong is one of the central financial hubs on the planet. Data based on the number of people working in the finance sector in Hong Kong is presented in this data set.

COMMON ERRORS

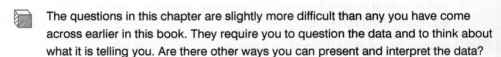 The questions in this chapter are slightly more difficult than any you have come across earlier in this book. They require you to question the data and to think about what it is telling you. Are there other ways you can present and interpret the data?

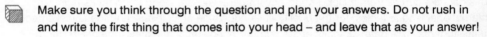 Make sure you think through the question and plan your answers. Do not rush in and write the first thing that comes into your head – and leave that as your answer!

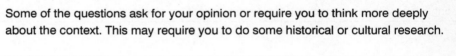 Some of the questions ask for your opinion or require you to think more deeply about the context. This may require you to do some historical or cultural research.

Hong Kong is a bustling cosmopolitan metropolis, a place where East meets West. This highly populated region of Asia houses a multitude of industry sectors, with a thriving labour market. This case study presents data in several forms, based on the leading industry sectors in the Sai Kung district of Hong Kong in 2018. This data has been modified from its original form. Modifications include shortening the data set, so it can fit on the page adequately. In addition, the graphical outputs in Figures 8.1–8.3 are based on the data presented in Table 8.3. The cross (x) in the blue box in Figure 8.3 indicates the mean value for this data.

Table 8.3 Industry sections and number of persons engaged in the Sai Kung district, Hong Kong, 2018

Year	District		
2018	Sai Kung		
	Number of persons engaged		
Industry section	**Male**	**Female**	**Total**
Mining and quarrying	33	5	38
Manufacturing	2,183	692	2,875
Electricity and gas supply, and waste management	118	47	165
Construction sites (manual workers only)	9,891	2,028	11,919
Import/export, wholesale and retail trades	4,754	6,355	11,109
Financing and insurance	1,354	1,011	2,365
Professional, scientific and technical services	768	638	1,406
Social and personal services	6,594	10,309	16,903
Education	4,448	6,174	10,622
Human health and social work services	858	3,049	3,907
Arts, entertainment and recreation	635	401	1,036

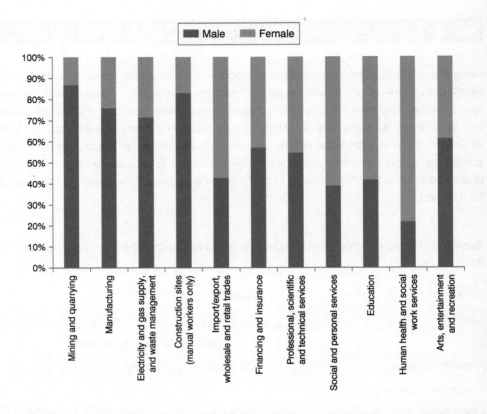

Figure 8.1 Stacked bar graph of Males and Females engaged in industry, Sai Kung district, Hong Kong, 2018

Answer the following questions related to this data.

1 Which Industry section had the highest proportion of Female workers in the Sai Kung district in 2018?

 >

 >

 >

 >

 >

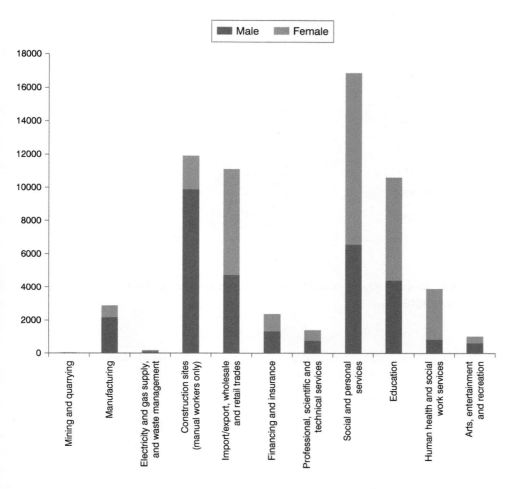

Figure 8.2 Stacked bar graph of persons engaged in industry, Sai Kung district, Hong Kong, 2018

2   Which Industry section had the highest number of Female workers in the Sai Kung district in 2018?

   >

   >

   >

   >

   >

113

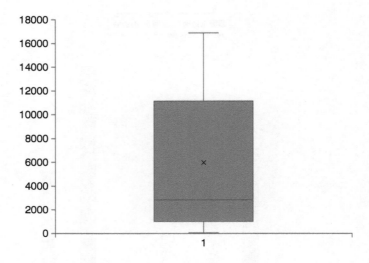

Figure 8.3 Box and whisker plot of Males engaged in industry, Sai Kung district, Hong Kong, 2018

3 Which Industry section had the lowest proportion of Female workers in the Sai Kung district in 2018?

>

>

>

>

>

4 Which Industry section had the lowest number of Female workers in the Sai Kung district in 2018?

>

>

>

>

>

5 Which Industry section had the proportion closest to 50% for Males and 50% for Females in the Sai Kung district in 2018?

>

>

>

>

>

6 Are the numbers and proportions of Females and Males listed in the Industry sections for Sai Kung district in 2018 surprising? Are they what you would expect? Explain your answer.

>

>

>

>

>

7 Would you say there is gender equality across the Industry section for the Sai Kung district in 2018? What reasons might there be for your answer?

>

>

>

>

>

8 What are the median and mean values for Males engaged in industry in the Sai Kung district in 2018? (The cross on the box-and-whisker plot in Figure 8.3 denotes the mean value.)

>

>

>

>

>

9 What is the interquartile range for Males engaged in industry in the Sai Kung district in 2018?

>

>

>

>

>

10 Which Industry section had the highest total number of workers in the Sai Kung district in 2018?

>

>

>

>

>

Answers can be found at the back of the book. Test your knowledge further by answering multiple choice questions at https://study.sagepub.com/jones

Hong Kong is world renowned for having an excellent education system, from primary school right through to higher education. Education institutes, as well as national student literacy and numeracy rates, often rank highly in world league tables. This case study focuses on the education sector in multiple regions of Hong Kong, in 2000. The data is arranged by sex, by region, or in some cases by both, and includes graphs presented with numbers and percentages. The data has been downloaded and modified from the Hong Kong government website, which includes the shortened version of the relevant data set. Graphs (Figures 8.4–8.6) have been generated using data presented in Table 8.4.

Table 8.4 Number of establishments and Number of persons Engaged in Education, by District, Hong Kong, 2000

Year	Industry section			
2000	Education			
			Number of persons engaged	
District	**Number of establishments**	**Male**	**Female**	**Total**
Central and Western	240	4,617	6,068	10,685
Wan Chai	306	3,772	4,721	8,493
Eastern	249	1,716	4,092	5,808
Southern	98	1,774	2,605	4,379
Yau Tsim Mong	342	3,960	4,766	8,726
Sham Shui Po	153	3,179	4,126	7,305
Kowloon City	221	2,938	5,609	8,547
Wong Tai Sin	139	1,264	3,310	4,574
Kwun Tong	185	1,894	3,996	5,890
Kwai Tsing	132	1,661	3,206	4,867
Tsuen Wan	120	1,078	2,106	3,184
Tuen Mun	159	2,060	3,766	5,826
Yuen Long	187	1,997	3,627	5,624
North	117	992	2,139	3,131
Tai Po	113	1,965	3,590	5,555
Sha Tin	256	5,709	8,247	13,956
Sai Kung	106	2,533	3,127	5,660
Islands	68	254	631	885
Total	**3,191**	**43,363**	**69,732**	**113,095**

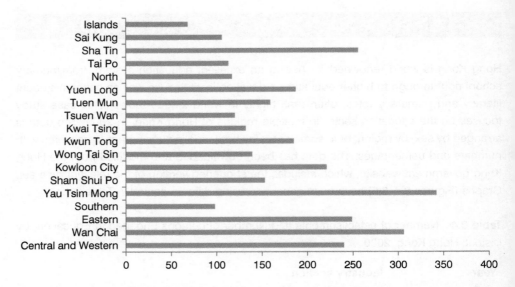

Figure 8.4 Bar graph of establishments in Education, Hong Kong, 2000

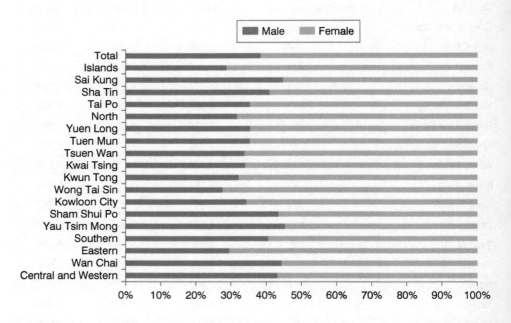

Figure 8.5 Stacked bar graph of the percentage of Males and Females engaged in Education, Hong Kong, 2000

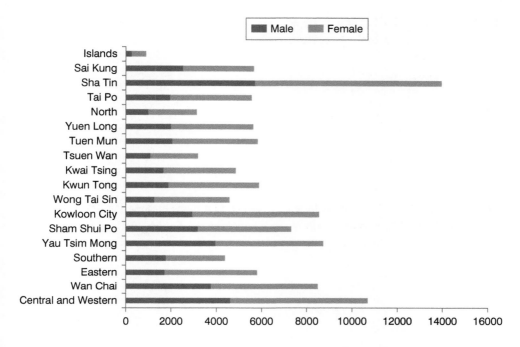

Figure 8.6 Stacked bar graph of the total number of Males and Females engaged in Education, Hong Kong, 2000

Answer the following questions related to this data.

11 How would you describe the proportions of Males and Females engaged in Education in Hong Kong in 2000?

>

>

>

>

>

12 What reasons could there be for the trend you have described in question 11?

>

>

>

>

>

13 How do you think funding of the educational establishments in Hong Kong was determined by region in 2000? Use any of the graphs in this case study to support your answer.

>

>

>

>

>

14 Do you think the Total percentage of Males and Females engaged in Education in Hong Kong in 2000 is an accurate representation of the proportions of Males and Females in each region of Hong Kong? Explain your answer.

>

>

>

>

>

15 Is there any pattern between the Number of establishments in Education, and the Number of Males and Females engaged in Education in Hong Kong in 2000?

>

>

>

>

>

Answers can be found at the back of the book. Test your knowledge further by answering multiple choice questions at https://study.sagepub.com/jones

Hong Kong plays host to a booming financial sector, employing thousands of citizens from the local region. The final case study in this chapter presents data on the Financing and insurance industry, from multiple districts in Hong Kong, between 2008 and 2009. These time points were specifically chosen, in relation to a world event that had a profound effect on the financing and industry sector at this time. The data is presented as numbers and percentages of Males and Females working in this industry sector across Hong Kong. Again, data was downloaded and modified from the Hong Kong government website. Data in Table 8.5 was used to generate the accompanying graphs and Table 8.6, to build up a story of the selected industry in this region of Asia.

Table 8.5 Number of establishments and Number of persons engaged in the Financing and insurance industry, by District, Hong Kong, 2008 and 2009

Industry section: Financing and insurance			**Number of persons engaged**		
Year	**District**	**Number of establishments**	**Male**	**Female**	**Total**
2008	Central and Western	3,682	47,583	48,611	96,194
2008	Wan Chai	3,173	11,093	12,492	23,585
2008	Eastern	1,630	7,522	10,216	17,738
2008	Southern	223	434	493	927
2008	Yau Tsim Mong	2,868	7,978	10,898	18,876
2008	Sham Shui Po	599	1,199	1,743	2,942
2008	Kowloon City	505	1,049	1,478	2,527
2008	Wong Tai Sin	262	425	595	1,020
2008	Kwun Tong	558	5,494	4,643	10,137
2008	Kwai Tsing	280	683	1,095	1,778
2008	Tsuen Wan	356	729	947	1,676
2008	Tuen Mun	248	303	664	967
2008	Yuen Long	339	533	819	1,352
2008	North	247	479	586	1,065

(Continued)

121

Table 8.5 (Continued)

Industry section: Financing and insurance   **Number of persons engaged**

Year	District	Number of establishments	Male	Female	Total
2008	Tai Po	286	390	661	1,051
2008	Sha Tin	415	902	1,542	2,444
2008	Sai Kung	354	947	826	1,773
2008	Islands	80	191	176	367
2008	**Total**	**16,105**	**87,934**	**98,485**	**186,419**
2009	Central and Western	3,902	46,081	43,850	89,931
2009	Wan Chai	2,956	9,179	10,796	19,975
2009	Eastern	1,895	8,591	10,445	19,036
2009	Southern	163	229	359	588
2009	Yau Tsim Mong	2,987	10,079	12,614	22,693
2009	Sham Shui Po	543	939	1,458	2,397
2009	Kowloon City	567	1,274	1,451	2,725
2009	Wong Tai Sin	347	521	741	1,262
2009	Kwun Tong	695	5,157	5,639	10,796
2009	Kwai Tsing	354	712	834	1,546
2009	Tsuen Wan	361	645	898	1,543
2009	Tuen Mun	308	522	575	1,097
2009	Yuen Long	407	767	1,137	1,904
2009	North	232	369	619	988
2009	Tai Po	304	589	633	1,222
2009	Sha Tin	455	1,393	1,754	3,147
2009	Sai Kung	272	349	478	827
2009	Islands	116	187	309	496
2009	**Total**	**16,864**	**87,583**	**94,590**	**182,173**

Table 8.6 Percentage difference of persons engaged in the Financing and insurance industry in Hong Kong, 2008 and 2009

Industry section: Financing and insurance

District	2008	2009	Difference 2008 to 2009	Percentage difference 2008 to 2009
Central and Western	96,194	89,931	−6,263	−6.51
Wan Chai	23,585	19,975	−3,610	−15.31
Eastern	17,738	19,036	1,298	7.32
Southern	927	588	−339	−36.57
Yau Tsim Mong	18,876	22,693	3,817	20.22
Sham Shui Po	2,942	2,397	−545	−18.52
Kowloon City	2,527	2,725	198	7.84
Wong Tai Sin	1,020	1,262	242	23.73
Kwun Tong	10,137	10,796	659	6.50
Kwai Tsing	1,778	1,546	−232	−13.05
Tsuen Wan	1,676	1,543	−133	−7.94
Tuen Mun	967	1,097	130	13.44
Yuen Long	1,352	1,904	552	40.83
North	1,065	988	−77	−7.23
Tai Po	1,051	1,222	171	16.27
Sha Tin	2,444	3,147	703	28.76
Sai Kung	1,773	827	−946	−53.36
Islands	367	496	129	35.15

16 Are Males and Females equally represented in the Financing and insurance industry in Hong Kong, in 2008 and 2009?

>

>

>

>

>

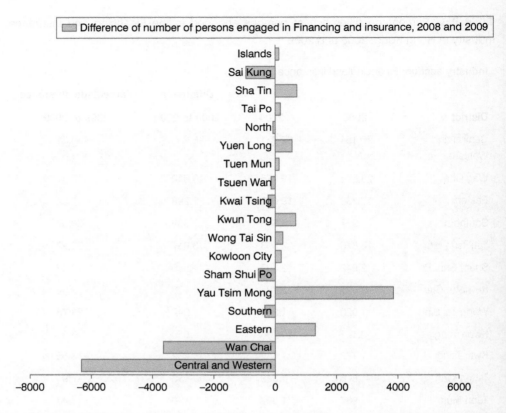

Figure 8.7 Bar graph of Number of persons engaged in the Financing and insurance industry, Hong Kong, 2008–2009

17 Which Districts in Hong Kong, between 2008 and 2009, saw the biggest percentage decreases in staff numbers? Is there any link with the Number of establishments in these districts?

>

>

>

>

>

18 What are the advantages and disadvantages of using numbers and percentages when reporting values?

>

>

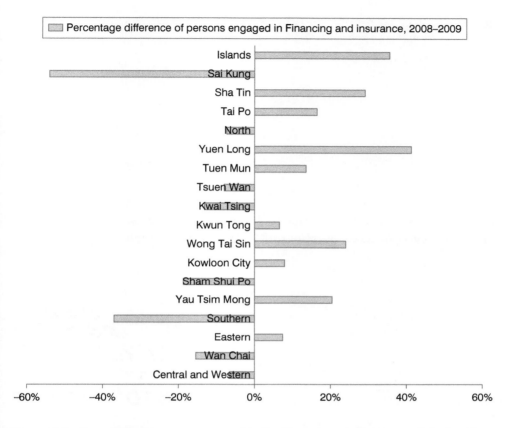

Figure 8.8 Bar graph of persons engaged in the Financing and insurance industry, Hong Kong, 2008–2009

>

>

>

19 Calculate the population mean and standard deviation for Males and Females in the Financing and insurance industry in Hong Kong in 2008. Feel free to use an online standard deviation calculator (for example, https://www.calculator.net/standard-deviation-calculator.html).

>

>

>

>

>

20 How do these values compare to the 2009 values? Feel free to use an online standard deviation calculator (for example, https://www.calculator.net/standard-deviation-calculator.html).

>

>

>

>

>

Answers can be found at the back of the book. Test your knowledge further by answering multiple choice questions at https://study.sagepub.com/jones

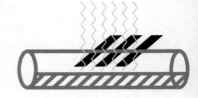

TIPS TO REMEMBER

Feature-spotting questions are extremely useful general questions to ask yourself about any data you come across (with some added questions to ask yourself!):

1 What are the main features I can see in the graph or table?

2 What does this mean?

3 What other details are useful for understanding the variable or answering any questions that I may have?

4 What other questions do I have?

5 Think about confounding factors/variables .Are there other things that could be influencing the data you have? These could be historical, cultural or political factors, to name a few.

6 Make sure you plan your answers carefully. Don't immediately write the first thing that comes into your head – and leave that as your answer!

9
OBSERVATIONAL STUDIES AND EXPERIMENTS

Skills in this chapter

Data is all around us! But where did it come from? How was it generated? Why was it generated? Can we trust it? What do we want to say about it? The next part of our data journey will help to answer these important questions, and will require us to think about how we view the social world, and how this determines our views on what we are willing to accept as reliable and valid data. The methods used to generate and analyse data should be selected after the research questions have been identified. This chapter will explore some of these concepts, and briefly touches on the philosophy of knowledge.

Knowledge, validity and reliability

Knowledge exists in many forms and can be viewed as the basis of humanity's desire to develop expertise in a variety of subjects. The approaches used to generate and analyse the data needed to reaffirm and create new knowledge are partly determined by the discipline in which it sits. For example, chemists and biologists often follow the scientific method in order to generate the data to support or refute a hypothesis they have come up with to solve a problem or gap in knowledge. They often choose experimental methods, whereby they control certain conditions and manipulate others to generate quantitative data. In using this approach, these scientists are able to replicate experiments to help improve the reliability of their results. In addition, if they are able to see similar patterns over time, scientists may be able to generalise these results externally (called external validity).

Sociologists view the social world in many ways, and this can determine the approaches they take to generate data. For example, social scientists may refute the use of the scientific method to measure gender interactions in the classroom, arguing that qualitative approaches are more appropriate to create rich and detailed data. For them, observation data taken over time (longitudinally, for example), could provide much more insightful information than using quantitative approaches (a class survey, for example).

Reliability explains the degree to which a research instrument (a stopwatch measuring time, for example) measures a given variable consistently every time it is used under the same conditions with the same subjects. Reliability usually refers to data and not necessarily to measurement instruments. From different perspectives or approaches, researchers can evaluate the extent to which their instruments provide reliable data.

Validity refers to the accuracy of research data. A researcher's data can be said to be valid if the results of the study measurement process are accurate. That is, a measurement instrument is valid to the degree that it measures what it is supposed to measure. There are different types of validity. *Internal* validity refers to whether there is a causal relationship between the variable being changed (called the independent or explanatory variable) and the variable being measured (called the dependent or response viable). *External* validity refers to how well one can generalise research results to other settings, programmes, persons, places, etc.

These concepts can be difficult to understand at first, but they become easier to understand when supplemented with examples. These will be provided in the following sub-headings.

Experiments

An experiment involves a researcher changing or manipulating certain conditions, to see what effect this has on a response variable. In certain kinds of experiments (laboratory based, for example), the researcher can determine which groups receive the treatment, and they can also include the use of control groups. The inclusion of a control group, who do not receive the treatment or intervention, can help in making useful comparisons between groups. The main aim of an experiment is to establish a causal relationship; for example, whether a change in one variable leads to and causes a change in the response variable.

Experiments can also include the random allocation of a treatment to groups (often called randomised control trials), to try to mitigate any bias in the data.

Some studies also include a blinding process, where the researchers or participants do not know which group received the treatment. This can help prevent any biases occurring when the researchers interpret the results.

An example of an experiment could come from the activities of a research group interested in childhood development of alcoholic behaviours. The researchers develop an intervention to help support children of alcoholic parents, to try to prevent the children also developing alcoholic tendencies. In their study, they recruit a group of families, in which one of the parents has identified as being an alcoholic. The researchers then randomly allocate families into two different groups. Those in group 1 receive a 16-week course to help educate parents and prevent their children from developing alcoholic behaviours, while those in group 2 receive no training over the 16 weeks (control group). All families are followed up over a 3-year period, with the children being asked to fill in a survey, measuring whether they have developed alcoholic behaviours (the response variable).

In this study the features which make it an experiment are:

- the implementation of an intervention, which is the treatment (or explanatory variable)
- this is also an example of a randomised control trial since the researchers used a random allocation process to determine which family groups received the treatment.

If, after 3 years, the researchers find that the children in intervention groups show little or no signs of developing alcoholic behaviours, then they might conclude that the intervention caused this change in behaviour. In addition, if the control group did show a marked increase in their children developing alcoholic behaviours, the researchers could be in a position to claim that their intervention helped to prevent this from happening in the intervention group.

There are two important points to mention with this example. First, ethical considerations are very important, especially in research of this kind. Research (including observational studies) often requires the researchers to apply for ethical clearance from an ethics board. This ensures that the subjects involved in the study (which can include animals, for example) are protected against any potential physical or psychological harm, and that the appropriate safeguards are put in place if any harm does occur.

Second, experiments rely on the researcher controlling certain variables and then manipulating others, to see if there are any causal relationships present. In research like the example described above, controlling certain variables can be difficult, if not impossible. We call these confounding variables. For example, in the experiment above, the researchers did not control for the age of participants (having been unable to select families that have similar ages to each other). This could have led to the inclusion of families whose parents were, say, 85 and 86, and parents who were 18 and 19. You could then argue that this variable (different ages of parents) could explain any differences seen in children's alcoholic behaviours after the 3-year follow up. For example, the older parents may have been stricter and clung to traditional values, so their children showed lower levels of developing alcoholic behaviours, while younger parents may have been less strict. These issues will directly influence the internal and external validity of the experiment. However, the process of randomly allocating the treatment to families in this example could help to mitigate any confounding variables identified.

Observational studies

Observational studies often involve a researcher observing particular groups of people, animals, or objects of interest, at a single point in time or over a longer period. These studies can include the collection of data using qualitative (e.g. group interviews) or quantitative approaches (e.g. surveys), or mixed methods (possibly a combination of qualitative and quantitative methods). The researcher might be involved with measuring changes to a response variable over time, looking for potential causal relationships, or might be interested in collecting data over time to look for common themes.

Observational studies can be *cross-sectional*, providing a snapshot of data at a point in time from a research subject of interest, or *longitudinal*, following the research subjects over a period of time, which can include data collection at multiple time points. Cohort studies, such as the Caerphilly Cohort Study, in Wales, or the Dunedin Study, in New Zealand, are examples of longitudinal studies.

An example of an observational study could include a social science researcher investigating identity and community in homeless people in a certain area. The researcher might even go undercover and pose as a homeless person themselves to gain insider research perspectives (think about the ethical considerations to consider with this example). Their data might include observational data of the physical environment or could include transcribed recordings from conversations with other homeless people. The researcher could then look for common themes, perhaps from common words used to describe an event, mentioned in the recordings. This type of research is an example of a qualitative approach. It could also be longitudinal, if the researcher decided to take snapshots of data across a period of time.

As with experiments, ethical considerations need to be explored, as well as clearance provided from an ethics committee.

Confounding variables are also a problem with observational studies, and some would argue that they are more problematic since variables that could have an effect on any causal relationships proposed are not being controlled. For these reasons, observational studies are useful for identifying possible causes of effects, but they cannot reliably establish causation.

Types of question in this chapter

Questions in this chapter will test your knowledge and understanding of experiments and observational studies. Development of your critical thinking skills will also feature prominently in this chapter. Deciding on the best approaches to take when answering research questions and deciding how you view the world around you will help to deepen your understanding of how social scientists think, as well as other scientists.

Resources to support this chapter

The case studies selected for this chapter come from two journal papers that were produced from the data generated by the Caerphilly and Speedwell cohort study, and one paper from an experiment based on light levels in restaurants and whether this can influence a person's choice to eat healthy food (see below).

Case study	Content	Theme	Country and date downloaded
9.1	Sex and death: are they related? Findings from the Caerphilly cohort study	Epidemiological studies	UK, 12 April 2019

https://www.bmj.com/content/315/7123/1641.full

This resource comes from a longitudinal cohort study, conducted in Caerphilly in South Wales, which started in 1979, lasting over 30 years. The research team investigated over 2,500 men in the Caerphilly region, looking at how people's lifestyles affect their health. This specific study looked at whether there is a link between orgasm frequency and mortality.

(Continued)

Case study	Content	Theme	Country and date downloaded
9.2	Traffic noise and cardiovascular risk: the Caerphilly and Speedwell studies, third phase 10-year follow-up	Epidemiological studies	UK, 12 April 2019

https://www.tandfonline.com/doi/abs/10.1080/00039899909602261

This resource comes from a longitudinal cohort study, conducted in Caerphilly in South Wales, which started in 1979, lasting over 30 years. The research team investigated over 2,500 men in the Caerphilly region, looking at how people's lifestyles affect their health. This specific paper looked at whether there is a between traffic noise and heart disease.

Case study	Content	Theme	Country and date downloaded
9.3	Shining light on atmospherics: how ambient light influences food choices	Marketing research	USA, 12 April 2019

https://journals.sagepub.com/doi/full/10.1509/jmr.14.0115

This resource looks at whether there is a link between light levels and the healthy food choices people make.

COMMON ERRORS

 Students are often taught that experiments are the best way to generate data, helping to keep the researcher as objective as possible, and helping to minimise bias in data interpretation. It really is not as simple as this and depends more on the type of research you are undertaking, the research questions posed, and, more importantly, the way you view the social world.

 Be careful how you interpret results from a study, paying close attention to the phrases you use to describe any potential causal relationships identified.

 A good researcher always comments on the strengths and weaknesses of any research carried out. This could include issues linked to ethics, as well as the validity and reliability of the methods and results.

This case study comes from a longitudinal cohort study, conducted in Caerphilly in South Wales, which started in 1979, lasting over 30 years. The research team investigated over 2500 men in the Caerphilly region, looking at how people's lifestyles affect their health. The study inspired over 400 journal papers, leading to further research around the world. The paper we are concerned with here, by Smith et al., reports on whether sex and death are related. Have a look at the variables measured below and answer the questions that follow.

This part of the Caerphilly cohort study involved the following:

Objective: To examine the relation between frequency of orgasm and mortality.

Study design: Cohort study with a 10-year follow up.

Setting: The town of Caerphilly, South Wales, and five adjacent villages.

Subjects: 918 men aged 45–59 at time of recruitment between 1979 and 1983.

Main outcome measures: All deaths and deaths from coronary heart disease.

Result: Mortality risk was 50% lower in the group with high orgasmic frequency than in the group with low orgasmic frequency.

Conclusion: Sexual activity seems to have a protective effect on men's health.

Key messages

- Sex and death are common variables in epidemiology, but the relation between them has not been studied in depth.
- In this cohort study, mortality risk was 50% lower in men with high frequency of orgasm than in men with low frequency of orgasm.
- These findings contrast with the view common to many cultures that the pleasure of sexual intercourse may be secured at the cost of vigour and well-being.
- If these findings are replicated, there are implications for health promotion programmes.

Answer the following questions related to this research.

1 Is this study an experiment or observational?

>

>

>

>

>

2 How do you think the research team collected data linked to the frequency of orgasms from the participants?

>

>

>

>

>

3 Why did the research team select only men?

>

>

>

>

>

4 Why did the researchers only select individuals aged 45–59?

>

>

>

>

>

5 Do you think the results in this paper are representative of people across all of Wales? Explain your answer.

>

>

>

>

>

6 Do you think the participants involved in the study reported their sexual activity honestly? Explain your answer.

>

>

>

>

>

7 Why do you think the results in this study are the opposite of what others have found?

>

>

>

>

>

8 The authors suggest that if the findings are replicated in other studies, then there are implications for health promotion programmes. Why are they making this claim?

>

>

>

>

>

9 Are there any ethical considerations associated with this research study?

>

>

>

>

>

10 What other factors could explain the findings reported here?

>

>

>

>

>

Answers can be found at the back of the book. Test your knowledge further by answering multiple choice questions at https://study.sagepub.com/jones

This case study comes from a longitudinal cohort study, conducted in Caerphilly in South Wales and England, which started in 1979, lasting over 30 years. The paper by Babisch et al. investigated whether traffic noise and cardiovascular risk are related. Have a look at the variables measured below and the design of the study, and answer the questions that follow.

Objective: To examine the relation between road traffic noise and heart disease.

Study design: Cohort study with a 10-year follow up.

Setting: The towns of Caerphilly (South Wales) and Speedwell (England).

Subjects: 2512 men (Caerphilly) and 2348 men (Speedwell) aged 45–59 at time of recruitment between 1979 and 1983.

Main outcome measures: The occurrence of a major heart disease event.

Result: The risk of heart disease increases when living near roads with high levels of traffic noise, for a longer period of time.

Answer the following questions related to this research.

11 What type of study is this – observational or experimental?

>

>

>

>

>

12 Explain your answer to question 11.

>

>

>

>

>

13 Is there any value in using two different locations for this study? Why?

>

>

>

>

>

14 What other factors could have caused changes in the main outcome measure?

>

>

>

>

>

Answers can be found at the back of the book. Test your knowledge further by answering multiple choice questions at https://study.sagepub.com/jones

This case study looks at whether there is a link between light levels and the healthy food choices people make. The paper created from this research, by Biswas et al., was published in the *Journal of Marketing* in February 2017. The authors emphasise the importance of this type of research by stating that retail atmospherics is emerging as a major competitive tool, especially in the restaurant industry. The overall ambience in the setting or environment could influence the consumer experience.

A study investigated the role of ambient light levels on consumers' food choices in a restaurant setting. Different locations of a restaurant chain were selected in a major city in the USA. Each restaurant was randomly allocated to either dim or bright ambient light levels. In each of the restaurants, the food choice and number of calories ordered by customers were recorded.

Some of the variables explored were:

- **Light**. The ambient light level in the restaurant (dim or bright)

- **Food**. The food choice (healthy or unhealthy)

- **Calories**. The number of calories in the food ordered.

Results. Consumers tend to choose less healthy food options when ambient lighting is dim.

Conclusion. This phenomenon occurs because ambient light luminance influences mental alertness, which in turn influences food choices.

Answer the following questions related to this research.

15 What type of study is this? Observational or experimental? Explain your answer.

>

>

>

>

>

16 If the people were aware that they were in a research study based on food choice, do you think it would influence what they select to eat? Why?

>

>

>

>

>

17 Are there any ethical considerations associated with this research study?

>

>

>

>

>

18 Do you think the location of the restaurant chains selected could have an impact on the data collected? Explain your answer.

>

>

>

>

>

19 The paper appears to be assuming that healthy food choices will result in lower numbers of calories. Would you agree with this assumption? Explain your answer.

>

>

>

>

>

20 Can the authors generalise these findings to other restaurants in the USA?

>

>

>

>

>

Answers can be found at the back of the book. Test your knowledge further by answering multiple choice questions at https://study.sagepub.com/jones

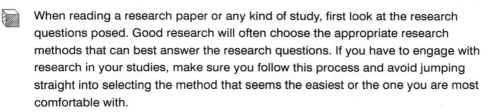

TIPS TO REMEMBER

When reading a research paper or any kind of study, first look at the research questions posed. Good research will often choose the appropriate research methods that can best answer the research questions. If you have to engage with research in your studies, make sure you follow this process and avoid jumping straight into selecting the method that seems the easiest or the one you are most comfortable with.

Remember there are always strengths and weaknesses to all research. Good papers often present recommendations for further research, which could be linked to the weaknesses of the paper. These provide opportunities to take their research further.

Do not fall into the trap of thinking that some research methods are better than others. All research methods have their strengths and weaknesses. Using mixed methods, involving quantitative and qualitative approaches, can provide you with powerful tools. However, this can also make it difficult to analyse and write up the research!

10
POLLS AND SURVEYS

Skills in this chapter

This final part of our data journey involves looking at polls and surveys, which are created for many different reasons and come in many forms. We are increasingly exposed to different ways of data being extracted from us, especially in digital form. For example, data collected from responses to a survey can be used to tailor emails that present you with specific offers from a retail store. Polls can be used to help predict the outcome of upcoming events, such as political elections. Polls tend to be based on one specific question, whereas surveys are larger, often including multiple questions and question types. In both forms of data collection, the results are often presented as percentages, proportions and mean values. There are also other considerations to reflect upon when evaluating the usefulness of polls and surveys, for example their validity and reliability – can the results be generalised to larger populations? This chapter will address these areas, followed by a series of case studies with questions to help you understand the topics covered.

Sampling

Presenting a survey or poll to a sample from a population of interest is often faster, cheaper and more practical than trying to get data from the whole population. Poll and survey reports should include the target population, sampling method, sample size, date and the exact questions asked.

Random sampling helps to avoid subjectivity in choosing participants, and allows for the calculation of sampling error. The larger the sample taken, the better, in terms of it being representative of the population of interest.

Sampling and non-sampling errors

Bias occurs when the data collected is consistently over-, or underestimated. A biased selection process is an inadequate selection or sampling process, which can lead to systematic biases.

Sampling errors occur as a result of taking a sample, affecting how representative it is of the population it came from. They also have the potential to be bigger in smaller sample sizes.

Non-sampling errors can be much larger than sampling errors and are always present. They are often impossible to correct for after the poll or survey has taken place. Any potential non-sampling errors must be minimised in the design phase of the poll or survey.

Types of non-sampling errors include the following:

Selection bias. The population sampled is not exactly the population of interest. For example, asking the readers of Vogue magazine (USA) for their opinions on same-sex marriage conducted via an online survey, then generalising the results to all North Americans.

Non-response bias. Choosing a certain group of people to be surveyed who do not respond, for example because they refuse to participate or are unreachable.

Self-selection bias. This is where people choose to volunteer to take part in a poll or survey, rather than being randomly selected. For example, a TV show could present a poll question and ask viewers to ring up or fill in their response online. The viewers themselves decide to participate.

Question effects. Variations in wording can have an effect on responses. For example, compare:

'Do you think there is a lack of discipline in children? Do you believe in national conscription?'

with:

'Do you think children should be given the time and space to grow, and play with other children? Do you believe in national conscription?'

Interviewer effects. Different interviewers asking the same question can obtain different results. This may be down to the sex, race, or religion of the interviewer. For example, when different interviews asked male participants, 'How often do you feel depressed?', 21% of female interviewers got the response 'Often', as opposed to only 6% of male interviewers.

Behavioural considerations. People tend to answer questions in a way they consider socially desirable. For example, people are more likely to say they have never cheated on a partner, even when they have.

Transferring findings. This involves taking the data from one population and transferring the results to another. For example, Londoners' opinions may not be a good indication of the opinions of people from Edinburgh.

Survey-format effects. Examples are question order, survey layout, and whether an interview is carried out by phone, in person or by mail.

Types of question in this chapter

Questions in this chapter will focus on a series of case studies, generated from online sources. These questions will help to deepen your understanding of the potential pitfalls that surveys and polls can create. These case studies will also get you to think about the different types of sampling and non-sampling errors that can occur, because of poorly designed survey questions.

Resources to support this chapter

The case studies in this chapter come from online sources that reflect examples of current practices used to extract data from internet users (see below). With more and more of us spending a lot more time online, exposure to these polls and surveys is on the increase. These resources will help to improve your critical thinking skills in being able to evaluate what makes a good poll or survey.

Case study	Content	Theme	Country and date accessed
10.1	Poll of favourite movie – *Toy Story*	Popular culture	USA, 16 May 2019

https://www.buzzfeed.com/ishabassi/which-of-these-movies-is-better-the-original-or-the-sequel

Polls are a popular way of generating data quickly, used by a variety of businesses and social media platforms. This resource presents you with a poll looking at Toy Story movies.

(Continued)

Case study	Content	Theme	Country and date accessed
10.2	The best online poll apps and social media polls in 2019	Social media polls	Global, 16 May 2019

https://zapier.com/blog/best-poll-apps/

This resource aims to provide the reader with tips on how to produce a good poll.

Case study	Content	Theme	Country and date accessed
10.3	2017 Census Test Report	Population studies	UK, 16 May 2019

https://www.ons.gov.uk/census/censustransformationprogramme/testingthecensus/2017test/2017censustestreport

This resource is an example of an official government document, which aims to generate data at a population level. Census data are often collected once every 4–5 years, to help provide important data to policy makers and resource allocators. This resource is an example of a test report, which would have been used to improve the main census survey.

COMMON ERRORS

Make sure you do not confuse sampling and non-sampling errors. Sampling errors occur because of taking a sample and are measurable. Non-sampling errors are usually unavoidable and very difficult to correct for after the data has been collected.

We are often bombarded with surveys asking for our opinion on our favourite music, food, holiday destination, movie, etc. Since many of us spend an increasing amount of time on social media platforms, and generally online surfing the internet, we often find surveys popping up on these platforms and even in our web browsers. This case study includes a snippet of a survey asking participants to select the best *Toy Story* movie (Figure 10.1). The curent results from the question are also displayed (Figure 10.2).

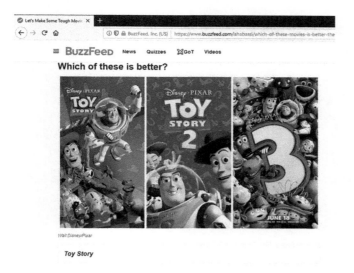

Figure 10.1 BuzzFeed survey (screenshot taken from a BuzzFeed survey on the best *Toy Story* movie. Source: BuzzFeed)

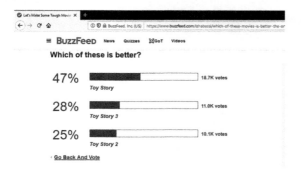

Figure 10.2 BuzzFeed survey results (screenshot taken from a BuzzFeed survey on the best *Toy Story* movie. Source: BuzzFeed)]

CASE STUDY 10.1

Answer the following questions related to this case study.

1 Do you think the question posed in the screenshot in Figure 10.1 is a good one?

>

>

>

>

>

2 How could this question be improved?

>

>

>

>

>

3 Are there any potential non-sampling errors with this part of the survey?

>

>

>

>

>

4 Do you think the results of this part of the survey (Figure 10.2) are a good way of presenting this data? Explain your answer.

>

>

>

>

>

5 Which additional pieces of information do you think would be useful to investigate the types of people who watch movies like *Toy Story*?

>

>

>

>

>

Answers can be found at the back of the book. Test your knowledge further by answering multiple choice questions at https://study.sagepub.com/jones

Marketing companies recognise the importance of survey and poll results, and as such offer their services to help companies create them (Figure 10.3). In addition, many online bloggers will offer their services in helping you select the best web-based polling apps, as well as how to create them in popular social media platforms. This case study explores some of the potential pitfalls of a society that could be experiencing survey fatigue!

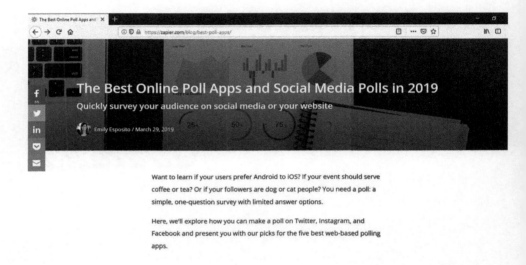

Figure 10.3 Zapier survey

Answer the following questions related to this case study.

6 The first paragraph in the screenshot in Figure 10.3 mentions limited answer options. Why do you think this is?

>

>

>

>

>

7   Check out this website online and scroll through the advice on the page. Is the information there useful? Try creating a poll of your own on a social media platform of your choice.

    >

    >

    >

    >

    >

Answers can be found at the back of the book. Test your knowledge further by answering multiple choice questions at https://study.sagepub.com/jones

Surveys can also attempt to collect data from a whole population, to help inform government-level policy changes, or perhaps influence government spending. A census is an example of a population-level survey, which attempts to collect data from every individual in a given country. This case study will explore the UK Census Test (which is not a population-wide survey), which was conducted in 2017. These types of activities are extremely important, and can act as a trial run for the bigger census surveys.

2017 Census Test Report

The 2017 Test involved 208,000 households in England and Wales. We tested a number of different 'treatments' to meet the test objectives.

Component one

Component one was conducted in seven local authority areas:

- Barnsley Metropolitan Borough Council
- Blackpool Borough Council
- Powys County Council (Montgomeryshire) (Cyngor Sir Powys)
- Sheffield City Council
- South Somerset District Council
- Southwark Council
- West Dorset District Council

We chose these areas because they include:

- a mix of rural and urban locations
- some areas with a substantial student population
- areas with a range of broadband coverage
- areas with concentrations of ethnic groups
- multilingual areas.

We sampled 100,000 households and used a split sample design. This tested the following 'treatments':

- a questionnaire with and without a sexual orientation question

- non-response field follow-up starting 4 days and 10 days after Census Test Day
- low, medium and high hard-to-count groups across the local authorities.

17 Which of the following options best describes how you think of yourself?

⟳ This question is voluntary

☐ Heterosexual or Straight

☐ Gay or Lesbian

☐ Bisexual

☐ Other, write in

Figure 10.4 2017 Census Test Report: Question 17

Answer the following questions related to this case study.

8 The question posed in this case study (labelled Question 17; see Figure 10.4) is offered as a voluntary one. Why do you think they have done this?

>

>

>

>

>

9 Are there any non-sampling errors to consider for this question, and the associated answers generated from it?

>

>

>

>

>

10 This case study describes why the areas they sampled from were selected. Comment on these selection criteria. Why do you think they were used?

>

>

>

>

>

11 The survey team mentioned their use of 'treatments' as part of the Census Test. Does this sound like an observational or experimental design?

>

>

>

>

>

12 Do you think the sample size for this test is sufficiently large enough to minimise sampling errors?

>

>

>

>

>

Answers can be found at the back of the book. Test your knowledge further by answering multiple choice questions at https://study.sagepub.com/jones

TIPS TO REMEMBER

 Make sure you do not confuse sampling and non-sampling errors.

 Sampling errors occur because of taking a sample and are measurable.

Non-sampling errors are usually unavoidable and very difficult to correct for after the data has been collected.

11

HOW WILL I USE STATISTICS?

Well done: you've made it to the end of the book! I hope that you are feeling more confident about looking at data, asking questions about it, and thinking of different ways it can be presented to tell a multiple stories.

If you have completed all the exercises in this book, along with the associated digital multiple-choice questions, you are well on your way to being able to start looking at more advanced types of data analysis and statistical concepts. Inferential statistics is usually the next step beyond this book, which involves looking at statistical and practical significance. Your studies may also take you into the realms of probability, and concepts related to randomness, which can have extremely useful applications.

Nurturing the skills you have developed in this book will give you the edge in any educational or career path you decide to embark on.

Remember that the data sets used in this book are in CSV (comma separated values) or Excel format, which means you will be able to continue exploring them using other software such as SPSS, R or iNZight. For example, you may want to look into using inferential statistical techniques. A good book for this would be Andy Field's *Discovering Statistics using IBM SPSS Statistics* (Sage, 5th edition). Just a word of caution: some of the data sets used in this book are large, so be patient when waiting for them to download. Everything you have seen in this book is sourced from real data, from across the globe.

Good luck! All the best for your future data journey.

ANSWERS

Chapter 2: Essential Numeracy

1   Birmingham
2   143 + 54 = 197
3   3
4   Blaenau Gwent
5   Bristol
6   The Licensing authorities could have different population sizes; that is, the larger Licensing authorities would have more business owners wanting to apply for alcohol licences, to serve the potentially higher demand for alcohol.
7   54 – 49 = 5
8   It would increase the number by 5 to 271 + 5 = 276.
9   57 + 3 = 60
10 Applied = 52 + 143 = 195, Granted = 52 + 124 = 176, Refused = 0 + 9 = 9
11 Hotels (i.e. not referring to the subdivisions of the larger premises types)
12 1832 + 711 = 2543
13 Premises type not reported (i.e. not referring to the subdivisions of the larger premises types)
14 'Supermarket and store type not reported'
15 749 + 50 = 799
16 2
17 22
18 Bothamley Park
19 Park no. 5
20 Bothamley Park
21 806,008.8 + 251,365.2 = 1,057,374 m^2
22 13,484.11572 – 168.6246326 = 13,315.4910874 m
23 Parks 5 and 6. These are Controlled areas.
24 1
25 12

Chapter 3: Percentages, Decimal Points and Inequalities

1 (901/7503) × 100 = 12%

2 (782/1815) × 100 = 43%

3 (307 + 41/7503) × 100 = 5%

4 (31/41) × 100 = 75.6%

5 (901 + 908/6013) × 100 = 30.08%

6 (26/7503 + 26) × 100 = 0.35%. The Total number of offenders would increase by 26 to 7529.

7 The 21+ age group, (4557 / 7503) × 100 = 61%

8 The over-3-months and up-to-4-months Length of sentence group, (1815/7503) × 100 = 24%

9 As age increases, so does the number of offenders receiving a custodial sentence.

10 As the Length of sentence increases, the number of offenders generally decreases.

11 (38/(27+38+68)) × 100 = 28.57%

12 (38/(27+38+68+77)) × 100 = 18.10%

13 Snacks

14 (5 × 169/(5×169 + 10×168 + 7×161)) × 100
= (845/845+1680+1127) × 100 = 23.14%

15 (5 × 169/(5×169+5×168+7×161)) × 100
= (845/845+840+1127) × 100 = 30.05%

16 1190.63 m^2

17 (2/24) × 100 = 8.333%

18 (8/24) × 100 = 33.333%

19 (10/24) × 100 = 41.667%

20 33.333 + 41.667 = 75.00%

21 $p > 80,000$

22 5

23 $P \geq 80,000$

24 0.12 – no

0.8 – no

1.5 – no

0.05 – yes

1.7 – no

0.03 – yes

25 0.05 would no longer agree with the inequality statement.

Chapter 4: Ratios and Proportions

1   8

2   8/20 (no need to simplify)

3   2

4   2/20 (no need to simplify)

5   5:20 = 1:4

6

Authority	2006/07	2007/08	2008/09	Total obese 10–11-year-olds
Manchester	11,400	10,950	11,300	33,650
Newcastle upon Tyne	9,585	9,360	9,855	28,800

The table should be completed using the following method. Multiply each percentage recorded in Table 4.2 by 50,000 for Manchester and by 45,000 for Newcastle upon Tyne. For example, the obesity rate for 10–11-year-olds in 2006/07 in Manchester is 22.8%. So the number required is 22.8% × 50,000 = 11,400.

7   11,400 / 33,650

8   28,800

9   9,360 / 28,800

10   219/640 = 200/600 = 2/6 = 1/3. As a ratio, this is 1:3.

11   Highest, Richmond upon Thames (64.1%); lowest, Barking and Dagenham (44.8%)

12   64.1 – 44.8 = 19.3%

13   Richmond upon Thames (64.1%), 70,000 × 64.1% = 44,870; Barking and Dagenham (44.8%), 70,000 × 44.8% = 31,360.

14   Richmond upon Thames, 70,000 – 44,870 = 25,130; Barking and Dagenham, 70,000 – 31,360 = 38,640.

15

Authority	Belonged	Didn't belong	Total
Richmond upon Thames	44,870	25,130	70,000
Barking and Dagenham	31,360	38,640	70,000

16   38,640 / 70,000

17   44,870 / 70,000

18   38,640 : 31,360 or approximately 5:4

19   44,870 : 25,130 or approximately 9:5

20   We need to know sample or population size to assess the extent of the claims being made with the Percentage. For example, a high percentage of a small number is still a small number. Proportions give you a better idea of the total sample or population size being reported.

Chapter 5: Tables

1 False – Louisiana, Illinois and Alaska had decreases in population.
2 California
3 True
4

	2012	2013	2014	2015	2016	2017
Females (55% × total population for each year)	1,513,825.5	1,532,600.85	1,557,451.5	1,585,681.35	1,616,589.7	1,648,921.45
Males (45% × total population for each year)	1,238,584.5	1,253,946.15	1,274,278.5	1,297,375.65	1,322,664.3	1,349,117.55
Nevada total population	2,752,410	2,786,547	2,831,730	2,883,057	2,939,254	2,998,039

5 There are big differences in population size for the many States of the USA. Since the total US population is a combination of all these, population growth or decline will be influenced more by States with bigger populations (California, for example), and the States or territories with smaller populations (e.g. District of Columbia) would go unnoticed when just looking at the total national population.

6 In general, the larger a State's population the more resources it will receive from central government. Similarly, it will account for a greater proportion of votes in an election than States with smaller populations.

7 The data could be consistent in the number of decimal places used for the Length and Area values. The data could be arranged in some order, perhaps alphabetically (by Name of Ground) or in ascending or descending order of Length and/or Area.

8 Opening times of the parks, whether they have car parking facilities, whether dogs are welcome all year round (i.e. not just during autumn/winter months), what facilities are available in the parks (e.g. toilets for people), and whether there are coffee shops, showers and places to grab a bite to eat.

9 The words 'Good', 'Improvement necessary', and so on could be replaced with numbers or Icons (smiley face, sad face, etc.). Also, the data could be displayed graphically.

10 Utilising numbers takes up less space, could make it easier to collate the data, and would enable you to carry out other types of statistical tests. Using icons, or other forms of visual display, could make it easier to identify which farms need improvement. Graphical display of the data would enable the reader to spot patterns more easily than in table format.

11 The farms could use the data to help assess whether they need to make improvements in their milking practices. They could also use it to market their services, assuming they had been given favourable reviews, and make comparisons with other farms, to assess potential competition.

12 Percentages are useful for making comparisons with different groups of data, which have different sample sizes, for example.

13 They can be misleading, especially if there are small sample sizes or fewer data points. Small changes in data points over time could result in large percentage Changes, which could suggest that observed differences are bigger than they are.

14 It may run a news story that focuses on crimes perceived to be more serious to society. In addition, it may focus on crimes that saw the biggest percentage increases (or decreases, depending on whether it wanted to put a positive or negative spin on the news story). The news channel may be less inclined to use actual numbers of crimes committed, since these are relatively small in Table 5.4, and may not have the same impact as using percentages.

15 The police forces in the Region could have changed the way they classify and report crime, with more crimes labelled as 'Other', which could explain the percentage increase. Without knowing what these other crimes are, it would be hard to make any definitive statements to explain these percentage increases.

Chapter 6: Introduction to Statistics

1 Categorical – Nominal
2 No. The description is brief and would benefit from additional information, perhaps in another column(s).
3 A more descriptive explanation of incidents would help Fire Service staff determine their urgency and the potential danger to human life. The time of the incident being reported would also be helpful, as well as the precise location of the incident.
4 Categorical – Nominal
5 Region 1
6 Continuous – Ratio

ANSWERS

7 Area (cm^2)
8 Categorical – Nominal
9 Glenelg
10 Yes! Split the column variables to represent one variable rather than two (i.e. separate out Description and Quantity). Clarification on some of the words in the table would help make the table clearer; for example, what does 'Reserve' mean in the Description/Quantity variable column?
11 Numeric – Discrete
12 Numeric – Discrete
13 2008
14 2007
15 Numeric – Continuous – Interval
16 Numeric – Continuous – Ratio
17 Diesel (all grades)
18 3,628,347,660 / 7,723,450,256 = 0.47 to 2 dp
19 Categorical – Nominal
20 3,628,347,660 – 4,334,718 = 3,624,012,942 litres

Chapter 7: Graphs

1 Between 300 and 400 tonnes
2 No, there is not a single data point that is far away from the others. There appears to be a split in a group of data points to the left and right of the graph.
3 Perhaps positively skewed, with the tail end of the data points spread towards the positive X-axis. It is hard to tell, since the data points are split into two areas.
4 No
5 2005 has the widest spread of data. 2011 has the greatest interquartile range.
6 2011
7 2005
8 Mostly decreasing
9 Potentially yes, the data points on the left-hand side of the graph are squashed together; this is a common problem with dot plots, when data points overprint. Also, with the type of waste included on the graph, there is some overprinting which could be distracting.
10 395–400
11 1133 – in this case best to use the table.
12 Visual data (i.e. in graph form) enables you to see the highest and lowest values more quickly, along with the shape and spread of the data. It does not, however, give you exact values, like a table does.

13 Region 1

14 Region 6

15 Perhaps Region 1 is larger. Alternatively, it has a higher population, which means people are more likely to need the Fire Service.

16 A bar graph would be useful in displaying the Situation found data. This could be problematic, however, since there are many categories in this variable, which would lead to a very large bar graph.

17 See above answer

18 Pie, stacked bar, stem-and-leaf (technically not a graph; however, it is another way of displaying data).

19 Graphs tell us a lot of information about the variables and the associated data they are displaying. Using the correct graph can tell us whether there is a relation between two variables – how one can potentially affect another. It can also tell us other things such as how variables change over time, or how different they are from each other (in the case of comparing categories in a bar graph, for example). Different graphs are available to enable us to tell a story, to get a certain message across to the reader.

20 Yes! From a practical perspective, it is often useful to think about the research aims and objectives, and any curiosities or questions you have about the area of interest. Then thinking about the type of data you want to collect will enable you to think more practically about whether the data is obtainable, ethical and achievable in the given period you have to do the research.

Chapter 8: Exploring Data and Descriptive Statistics

1 Human health and social work services

2 Social and personal services

3 Mining and quarrying

4 Mining and quarrying

5 Professional, scientific and technical services

6 The numbers and proportions of Males in manual labour Industry sections (Construction, Mining, Electricity and Gas and Manufacturing) are higher than for Females. The number and proportion of Females in areas such as Education, Human health and social work services, and Social and personal services are higher than for Males. These figures could reflect historic social norms, whereby Males tend to engage with manual-labour-type jobs more than Females, whereas Females tend to have higher representation in social work/education type jobs. There could be other

cultural considerations to consider, specific to Hong Kong and the Greater China region. Western influences could also have an impact on social norms and work-related stereotypes, linked to the history of Hong Kong being governed by the UK until 1997.

7   Building on answer to question 6: No, there is not gender equality among the listed Industry section in the Sai Kung District in Hong Kong in 2018. This could be due to historic roles undertaken by Males and Females in the region.

8   Median, 2875 (or 100 either side of this value); mean, 5668 (or 100 either side of this value).

9   The interquartile range is found by subtracting the lower quartile from the upper quartile: 11,109 – 1036 = 10,073 (or 200 either side of this value).

10 Social and personal services

11 Figure 8.5 shows that there are higher proportions of Females engaged in Education in Hong Kong in 2000. You could state the exact proportions of Males and Females (by calculating them from Table 8.4), and even compare individual regions to the Total value for Hong Kong in 2000. For example, the Districts of Eastern, Islands and Wong Tai Sin have the highest proportions of Females engaged in Education in 2000.

12 There are many reasons! More Males could be engaged in other Industry section. There might be a tradition in Hong Kong of more Females engaged in Education than Males. The definition of 'engaged in Education' is open to interpretation; for example, if we were to look at the proportions of males and females who are teachers, this could present us with data that looks different.

13 Areas with more education establishments are likely to receive higher levels of funding. For example, using Figure 8.4, Districts such as Yau Tsim Mong, Wan Chai and Sha Tin which have higher Numbers of establishments in education are likely to receive more funding to help run and maintain them.

14 The Total percentage of Males and Females engaged in Education is calculated as a percentage by combining all the values of Males and Females across all Districts. Districts that have larger overall numbers will have a higher level of weighting over the Total percentage of Males and Females. This could give a misleading representation and give perhaps a distorted view of District-level proportions.

15 The Districts with higher Numbers of establishments in Education appear to have more of an even distribution of Males and Females, for example Wan Chai, Yau Tsim Mong and Sha Tin.

16 It depends on whether you are comparing numbers or proportions. Looking at the *numbers* of Males and Females in each District for 2008 and 2009, they are pretty much equally represented. It really depends on what you classify as equally represented. From 2008 to 2009, there appears to be a decrease in the difference between Male and Female numbers in each District, with Eastern, Yau Tsim Mong, Sha Tin, Kowloon city and Wan Chai seeing the biggest difference in Males and

Females across both years. If you were to look at the *proportion* breakdown for each District, and then Total values, you are likely to get a very different picture! For example, Tuen Mun had a ratio of roughly 2:1 Males to Females in 2008, changing to almost 1:1 in 2009 (see Table 8.5). In terms of numbers, and when you compare this District to larger districts, they may or may not be important.

17 The biggest percentage decreases in staff numbers between 2008 and 2009 were in Central and Western, Wan Chai, Southern, Sham Shui Po and Kwai Tsing. Just looking at Table 8.5, these Districts include both high and low Numbers of establishments, so there does not appear to be any relation between the Numbers of establishments in a District and percentage decreases in staff.

18 This answer builds on several points made in the answer to question 16. The importance of using numbers and/or proportions can depend on the context or the thing you are measuring or reporting on. Numbers give you a good idea of the actual values – how many people were involved in a study, for example. However, using just numbers to compare between different groups, for example decreases or increases in staff numbers in this case study, can be difficult to achieve. Using proportions helps you to compare differences between groups, but without the numbers it might not give you an idea of how important these differences are. Using both numbers and proportions is useful to help you to decide if any differences identified are important, or worth reporting.

19 In 2008, for Males we have mean 4885.22 (2 dp) and sd 10,838.31 (2 dp); for Females, mean 5471.39 (2 dp) and sd 11,150.03 (2 dp).

20 In 2009, for Males, we have mean 4865.72 (2 dp) and sd 10,516.49 (2 dp); for Females, mean 5255.00 (2 dp) and sd 10,150.35 (2 dp). In 2009, the mean values are closer together, as are the standard deviations, which would support our conclusions made earlier about the differences between Males and Females decreasing from 2008 to 2009.

Chapter 9: Observational Studies and Experiments

1 Observational

2 Survey/questionnaire

3 There are differences between male and female physiology, with men being much more likely to develop heart disease later on in life. Just selecting men can help with ruling out sex as a possible explanatory variable.

4 Age is also a known factor in contributing to the incidence of heart disease in humans. Ensuring participants in the study are a similar age can help to rule out this variable as a possible explanation for the results.

5   The results are from an observational study and need to be considered carefully when making inferences about the population of Wales as a whole. If the results can be replicated in other studies which look at multiple groups of people across Wales with sufficiently large sample sizes, then this can help to increase our confidence in the research papers' findings as being applicable to all people in Wales.

6   This is hard to judge! One would hope so; however, when dealing with sensitive issues such as orgasm frequency, there is a chance that participants might not be as forthcoming as one would hope.

7   There are many potential answers to this question. There could be different people involved with other research studies investigating this phenomenon. Participants from different studies could have different ages, be from a culture that has different attitudes towards sex and the reporting of orgasms, and be exposed to different media platforms that could affect the frequency of orgasms. These different exposures and potentially different cultural attitudes towards sex and orgasms would need to be considered carefully, as well as their impacts on the development of heart disease.

8   This claim suggests that if others generate similar data to theirs, and they come to the same conclusion, then this can help strengthen the claim that sexual activity seems to have a protective effect on men's health. The message can then be used in health promotion programmes, potentially as a way to help improve men's heart health.

9   Yes! This study deals with sensitive issues, in relation to men reporting their orgasm frequency. For example, some men might be impotent, or choose not to orgasm. The research team would have had to both consider these ethical perspectives carefully, and obtain ethical clearance before conducting research of this nature.

10 Many other factors can explain coronary heart disease incidence, such as food intake, cholesterol levels, stress levels, and whether the participants are smokers. These factors would need to be controlled, or the results would need to be adjusted to take account of this.

11 Observational

12 The study involved following groups of participants over time, recording whether there was a major heart disease event. The researchers were not changing any conditions or manipulating variables.

13 Yes, using two different groups increases the sample size of the participants involved in the study. If a possible link or association is found between road traffic noise and major heart disease events, and this is found in different regions, it strengthens the case that there is a causal relationship present.

14 Similar answer to question 10. Many other factors can explain coronary heart disease incidence, such as food intake, cholesterol levels, stress levels, and whether the participants are smokers. These factors would need to be controlled, or the results would need to be adjusted to take account of this.

15 Experimental. The researchers were changing the Light variable.

16 Potentially yes! Look up the Hawthorne effect – this is a well-known potential downside when participants are aware that they are part of a study.

17 Potentially yes. Participants involved could have eating disorders or other associated psychological issues linked to eating food. This study could trigger or amplify these issues, so the researchers would need to be mindful of this, as well as providing counselling services for anyone affected in an adverse way.

18 Potentially yes. Certain US states might have different attitudes towards food choice, which could influence the amount of healthy food selected, irrespective of ambient lighting levels.

19 Not always, and this really depends on perception as to what counts as healthy food. For example, many would state that avocados are a healthy food choice; however, they have a high fat content (although referred to as healthy fat). Healthy foods do not necessarily mean fewer calories – it is much more complicated than that!

20 Care would need to be exercised if the authors decide to generalise these findings to all other restaurants in the USA. There are many confounding variables that can influence Food choice, other than ambient Light levels.

Chapter 10: Polls and Surveys

1 It really depends on how accurate the survey designers want to be, the type of data they want to collect and what uses they have for the data. The word 'better' is subjective and can mean different things to different people.

2 Using phrases such as 'Which one of these is your favourite?' or 'Which of these did you most enjoy?' could be an improvement.

3 Depending on how the question is distributed, there could be problems with selection bias, non-response bias, self-selection bias and survey-format effects.

4 Yes, since the percentages and sample sizes are included. A pie chart could be a good alternative to presenting this data as well.

5 Perhaps gender, age and whether the participant has any children.

6 Limiting answer options will reduce the type and amount of data collected, which could make it easier to analyse and present.

ANSWERS

7   Overall yes; apart from the constant pop-ups, a lot of the information is free, and you can create your own polls on social media platforms.

8   People might be uncomfortable stating their sexuality or how they think of themselves, so giving people the option to fill this in could help them feel safer and less pressured.

9   Since people are given the option to fill this in, non-response bias could be a problem. Question effects could also be a problem with this type of question.

10   The selection criteria suggest that the survey team were looking for a representative sample from the UK population, trying to capture people's data from a mix of rural and urban locations. They were also looking for areas with substantial student populations, which could reflect a specific objective of this Test Census.

11   This is an experimental design, because of the word treatment being used (i.e. they are changing the conditions in the experiment).

12   Yes, because a very large sample size was used.

GLOSSARY

Association   If the data points on the Y-axis increase in line with the X-axis, then we can say there is a positive association. If, however, the data points decrease on the Y-axis as the X-axis increases, then we say there is a negative association.

Behavioural considerations   People tend to answer questions in a way they consider socially desirable. For example, pregnant women being asked about their drinking habits are unlikely to admit to consuming alcohol, and people are more likely to say they have never cheated on a partner, even when they have.

Bias   This occurs when collected data is consistently over- or underestimated.

Biased selection   A biased selection process is an inadequate selection or sampling process, which can lead to systematic biases.

Categorical   Categorical data are often groups defined by words. They can be ordinal or nominal. Ordinal data have a natural order or hierarchy – for example, level of education (BSc, MSc, PhD) and income (low, middle, high). Nominal data have no order – for example, gender, type of car, occupation.

Confounding variables   These are other factors that could explain the results in our data. Experiments rely on the researcher controlling certain variables and then manipulating others, to see if there are any causal relationships present. Controlling certain variables can be difficult if not impossible – these are what we call confounding variables.

Descriptive statistics   These are used to describe the basic features of data in a data set. They provide simple summaries about the sample or population. With descriptive statistics, you are simply describing what the data is or what it shows.

Ethical considerations   These are important to consider when doing certain kinds of research. Most research (including observational studies) often requires the researchers to apply for ethical clearance from an ethics board. This ensures that the subjects involved in the study (e.g. animals) are protected against any potential harm (which could be physical or emotional), and the appropriate safeguards are put in place if any harm does occur.

Experiments   An experiment involves a researcher changing or manipulating certain conditions, to see what effect this has on a response variable. In certain kinds of experiments (laboratory based for example), the researcher can determine which groups receive the treatment, and they can also include the use of control groups. The inclusion of a control group, who do not receive the treatment or intervention, can help in making useful comparisons between groups. The main aim of an experiment is to establish a causal relationship, for example, whether a change in one variable leads to and causes a change in the response variable. Experiments can also include the random allocation of a treatment to groups (often called randomised control trials), to try to mitigate any bias in the data. Some studies also include a blinding process, where the researchers or participants do not know which group received the treatment. This can help prevent any biases occurring when the researchers interpret the results.

Explanatory variable   Sometimes called the *independent variable*, in the context of an experiment, this is the variable that is said to explain or predict any differences or changes we see in the response (or dependent) variable. The explanatory variable is the one an experimenter changes or modifies to see what effect it has on the response or dependent variable.

Interviewer effects   Different interviewers asking the same question can obtain different results. This could be due to the sex, race, or religion of the interviewer. For example, when asking male participants 'How often do you feel depressed?', a female interviewer found that 21% answered 'Often', while a male interviewer found 6% gave that answer.

Knowledge   Knowledge exists in many forms, and can be viewed as the basis of the humanities' desire to develop expertise in a variety of subjects. The approaches taken to generate and analyse the data needed to reaffirm and create new knowledge are partly determined by the discipline in which it sits. For example, chemists and biologists often follow the scientific method in order to generate data to support or refute a hypothesis they have come up with, to solve a problem or fill a gap in knowledge. They frequently choose experimental methods, whereby they control certain conditions and manipulate others to generate quantitative data. In using this approach, these scientists are able to replicate experiments to help improve the reliability of their results. In addition, if they are able to see similar patterns over time, scientists may be able to generalise these results externally (called external validity).

Mean   This is a statistical name for the ordinary, everyday average. It is where the dot plot balances. It can be calculated by adding up all the values of your data points and dividing by the number of data points you have.

Median   This is the middle value in your data.

Modality (or mode)   The modality of data refers to the most frequent points in your data set. If there were two clear modal values, and the data spread looked like a camel's back with two humps, we would say the data has a bi-modal distribution.

Non-response bias   This is due to choosing a certain group of people to be surveyed but who do not respond. For example, they may refuse to participate or be unreachable.

Non-sampling errors   These can be much larger than sampling errors and are always present. They are often impossible to correct for after the poll or survey has taken place. Any potential non-sampling errors must be minimised in the design phase of the poll or survey.

Numeric   Numeric data can be discrete or continuous, and are often measurements or counts. Discrete data are frequency counts – for example, number of pies sold, number of caps worn in a month. Continuous data are commonly measurements, and can be on either a ratio scale (e.g. weight of rabbits, height of trees, heart rate, distance, age) or interval scale (e.g. temperature, pH).

Observational   Observational studies often involve a researcher observing particular groups of people, or animals, or a thing of interest, at a single point in time or over a longer period. These studies can include the collection of data using qualitative (e.g. group interviews) or quantitative approaches (e.g. surveys), or mixed methods (which could be a combination of qualitative and quantitative methods). The researcher might be involved with measuring changes to a response variable over time, looking for potential causal relationships, or they might be interested in collecting data over time to look for common themes. Observational studies can be cross-sectional or longitudinal. Cross-sectional studies provide a snapshot of data at a point in time, from a research subject of interest, whereas longitudinal studies follow the research subjects over a period of time, which can include data collection at multiple time points. Cohort studies, such as the Caerphilly cohort study in Wales, or the Dunedin study in New Zealand, are examples of longitudinal studies.

Outliers   An outlier is a data point that appears to be far away from the rest of your data points. There is sometimes the temptation to remove outliers from a data set, which should be avoided! It is much better to ask questions about that data point. Why is it so far away from the others? Could there have been a mistake in reading the data? On the other hand, is the outlier of interest? Removing a perceived outlier can have big implications for any statistical analysis you perform on the data set, and might not give you the true answers.

Percentages   These enable the presentation of data on a scale between 0 and 100%. It is always useful to know what sample or population size a percentage is based on.

Polls and surveys These are created for many different reasons and come in many forms. We are increasingly exposed to different ways of data being extracted from us, especially in a digital format. For example, data collected from responses to a survey can be used to tailor emails that present you with specific offers from a retail store. Polls can be used to help predict the outcome of upcoming events, such as political elections. Polls tend to be based on one specific question, whereas surveys are larger, often including multiple questions and question types.

Population This includes all the data points of interest, which can come from a group of participants, people, plants, animals, etc. Often a subset of the population (a sample) is chosen to represent that population.

Proportions These can be presented as fractions or with a decimal point. For example, you could say $\frac{3}{4}$ or 0.75 (simplified from 15/20) voted for the first candidate while $\frac{1}{4}$ or 0.25 (simplified from 5/20) voted for the second.

Question effects Variations in wording can have an effect on responses. Compare 'Do you think there is a lack of discipline in children? Do you believe in national conscription?' and 'Do you think children should be given the time and space to grow, and play with other children? Do you believe in national conscription?'

Random sampling This helps to avoid subjectivity in choosing participants, and allows for the calculation of sampling error. The larger the sample taken, the better, in terms of it being representative of the population of interest.

Ratios These describe a relationship between two numbers, indicating how many times the first number contains the second. For example, if 20 people were asked to vote for one of two candidates, with 15 voting for the first candidate and 5 voting for the second, as a ratio this would be written as 3:1. Ratios and proportions are interchangeable, as are percentages and proportions.

Reliability This explains the degree to which a research instrument (e.g. a stopwatch measuring time) measures a given variable consistently every time it is used under the same conditions with the same subjects. Reliability usually refers to data and not necessarily to measurement instruments. From different perspectives or approaches, researchers can evaluate the extent to which their instruments provide reliable data.

Rounding Rounding to a certain number of decimal places (say, n) involves paying attention to the location of the decimal point. To round numbers accurately, count off n digits, right from the decimal point. Then you must decide what to do with the nth digit. Look at the number directly next to it, on the right. If it is 4 or less then the nth digit stays the same; if it is 5 or higher, you round the nth digit up by 1.

Sample A sample is a set of data collected from a population.

Sampling Presenting a survey or poll to a sample from a population of interest is often faster, cheaper and more practical than trying to get data from the whole population. Poll and survey reports should include the target population, sampling method, sample size, date and the exact questions asked.

Sampling errors These occur as a result of taking a sample – is it representative of the population it came from? They also have the potential to be bigger in smaller sample sizes.

Scatter This gets you to ask whether the data points are constant on a scatter plot. If you were to draw a straight line though the data points, are they close to the line? Alternatively, is there a lot of variation in data points around the line?

Selection bias This is the situation where the population sampled is not exactly the population of interest. It might occur, for example, if you asked readers of *Vogue* magazine (USA) for their opinions on same-sex marriage via an online survey, and then tried to generalise the results to all North Americans.

Self-selection bias This can occur when people choose to volunteer in a poll or survey, that is, they are not randomly selected. For example, a TV show could present a poll question and ask viewers to ring up or fill in their response online. Since the viewers decide themselves to participate (i.e. they are not selected), this can give rise to self-selection bias.

Skewness When looking at dot plots or histograms, how the data is shaped will tell you if it is positively skewed, negatively skewed or symmetrical. If the tail end of the data is located towards the positive X-axis values, then we call this positive skew. If the tail end of the data points is located towards the negative X-axis values, then we call this negative skew.

Standard deviation This can be thought of as the typical or average distance between the individual data points and the mean.

Strength of relationship If the data points on a scatter plot are close to a straight line drawn in such a way as to fit the data points as closely as possible, then we can say there is a strong relationship between the variables on the axis. If the data points are spread out and far away from each other, with no discernible pattern, then we say there is a weak relationship.

Survey-format effects Examples are question order, survey layout, and whether the interview was conducted by phone, in person or by mail.

Symmetry   Assessing the symmetry of your data points, which is best performed on dot plots or histograms, can be thought of as placing a line somewhere in the middle of the data points, and seeing if it mirrors each side. For example, if you could place a line in the middle of your data points on a piece of paper, and you folded it along the line you have drawn, would they roughly fit over each other?

Tables   These are used to present data, which can highlight interesting patterns or relationships.

Transferring findings   This involves taking the data from one population and transferring the results to another. For example, Londoners' opinions may not be a good indication of the opinions of people from Edinburgh.

Trend   The trend of data on a scatter plot can tell you whether there is a linear (straight-line) relationship, that is, whether you can roughly fit all the dots onto a straight line.

Validity   This refers to the accuracy of research data. A researcher's data can be said to be valid if the results of the study measurement process are accurate. That is, a measurement instrument is valid to the degree that it measures what it is supposed to measure. There are different types of validity. Internal validity refers to whether there is a causal relationship between the variable being changed (called the independent or explanatory variable) and the variable being measured (called the dependent or response variable). External validity refers to how well one can generalise research results to other settings, programmes, persons, places, etc.